Housing policy in the 1990s

The late 1980s saw a deluge of Conservative legislation designed to shake the foundations of post-war housing policy. What has this achieved so far and what are its effects during this decade and into the next century? Are we at a crossroads, able to make choices, or have we already passed the point of no return? Have profound underlying shifts in housing tenure and the balance of political forces in housing changed so rapidly that there may not be much choice left? The contributors to this book – some academics and some leading practitioners, but all experts in different aspects of the subject – have been challenged to provide some answers to these questions.

Housing Policy in the 1990s examines whether the 'enabling' local authority has really been 'disabled' by central government, whether housing associations can fulfil their new role as leading providers of social rented housing, whether building societies will still be able and willing to finance them, what sort of social and economic consequences the growth in home ownership will have, and whether the private rented sector can be revived. It provides critiques of government policies from the 'new right', from a race and gender perspective, and from the point of view of council tenants.

Housing Policy in the 1990s is essential reading for policy analysts, students and lecturers of social policy and housing courses, as well as those with an interest in urban studies, and economics.

Johnston Birchall, the editor, is Lecturer in Social and Public Policy at Brunel University.

Housing policy in the 1990s

Edited by Johnston Birchall

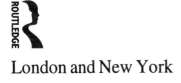

London and New York

First published in 1992
by Routledge
11 New Fetter Lane, London EC4P 4EE

Simultaneously published in the USA and Canada
by Routledge
a division of Routledge, Chapman and Hall, Inc.
29 West 35th Street, New York, NY 10001

Typeset by LaserScript Limited, Mitcham, Surrey
Printed and bound in Great Britain by
Biddles Ltd, Guildford and Kings Lynn

British Library Cataloguing in Publication Data
A catalogue record for this book is available from the British Library.

Library of Congress Cataloging in Publication Data
Housing policy in the 1990s / edited by Johnston Birchall.
    p.   cm.
    1. Housing policy – Great Britain   I. Birchall, Johnston.
. HD7333.A3H694   1992
    363.5′8′0941–dc20
                                                              92–10400
                                                                   CIP

ISBN 0–415–04358–1
    0–415–04359–X (pbk)

Contents

Contributors

Johnston Birchall is Lecturer in Social and Public Policy at Brunel University.

D.A. Coleman is Lecturer in Demography at the University of Oxford.

A.D.H. Crook is Reader in Town and Regional Planning at the University of Sheffield.

Norman Ginsburg is Principal Lecturer in Social Policy at South Bank University.

Mike Langstaff is Director of the Family Housing Association.

Stuart Lowe is Lecturer in Social Policy at the University of York.

Peter Malpass is Professor of Housing Policy at The University of the West of England, Bristol.

Douglas Smallwood is Manager, New Business Development Commercial Lending, at the Halifax Building Society.

Sophie Watson is Professor of Urban and Regional Planning at the University of Sydney, Australia.

Abbreviations of terms

ADC	Association of District Councils
ADP	Approved Development Programme
BCA	Basic Credit Approval
BES	Business Expansion Scheme
CGT	Capital Gains Tax
CRE	Community Relations Executive
DoE	Department of the Environment
GLC	Greater London Council
GRF	Grant Redemption Fund
HA	Housing Association
HAG	Housing Association Grant
HMOs	Houses in multiple occupation
HRA	Housing Revenue Account
HNI	Housing Needs Index
MIRAS	Mortgage Interest Relief at Source
NFHA	National Federation of Housing Associations
PLC	Public Limited Company
OPCS	Office of Population Censuses and Surveys
RDG	Revenue Deficit Grant
RPI	Retail Price Index
RSF	Rent Surplus Fund
RSG	Rate Support Grant
SCA	Special Credit Approval

Introduction

Policy analysts like to do more than keep up with their subject; they like to think they can, like the Old Testament prophets, predict something of what will happen, cut through the taken-for-granted half truths of political debate, even at times warn of impending disaster. The problem is that in a time of rapid change, when analysis is most urgently needed, analysts tend to lose the initiative, to become uncertain of what is happening, dependent on the unfolding of events to enable them to regain their certainties. For instance, it has been difficult to write accurately about current British housing policies, let alone to predict even the near future with any hope of success, while a Conservative government has been engaged on one of the most important and far reaching batches of housing legislation this century.

The word 'batches' is used deliberately, because analysts have agreed that it is the combination of the 1988 Housing Act and the 1989 Local Government and Housing Act which has been decisive; in some policy areas, such as the private rented sector, the policy thrust became clear in 1988, while in others, such as the future of council housing, we have simply had to wait for the legislation to be known in its entirety before being able to say anything much at all. The problem has been compounded by the peculiar nature of modern legislation, that it only becomes clear about a year on, when those areas sketched in by the politicians are filled out in detail by civil servants and ministers who have the capacity profoundly to alter not only the letter but in some cases the spirit of the legislation by explanatory guidelines, circulars and the like; the Tenants' Choice provisions of the 1988 Act come instantly to mind.

In consequence, some of the writers in this volume have been able to write sooner than others about the changes which are taking place. Some have had to wait until the main lines of the particular policy they are writing about become clear, and some have even then had to revise their chapters at the last moment, hoping that the pace of change had finally slowed and that the particular moment in which their contribution is frozen would prove decisive. One consequence of writing too soon is that statements made in good faith – about the direction Tenants' Choice would take, for instance – can become embarrassingly outdated. Another is that government initiatives can too soon be written off as failures; Housing Action Trusts have both failed dismally and, in the first few months of 1991, been resurrected, though in a modified form which invalidates some, but not all, of the critique developed in 1990.

Yet one consequence of not writing as soon as possible to offer accurate description, analysis, interpretation and prognosis – even prophecy of doom – on current housing policies and practices is to deprive students of the subject from being able quickly to grasp the essentials of the changes and the continuities which are taking us into the 1990s. And by students we mean all of us, from policy analysts to Institute of Housing students, to housing managers, tenant activists and anyone who has to live in and make sense of the complex world of housing policy. For this reason, the authors of the following chapters have decided to risk it; to write, sometimes with certainty, more often with an eye on some still unresolved questions, and always with hindsight, about some of the housing issues they regard as important for the future. The reader will find not a comprehensive, well-rounded textbook but a set of short essays which, as well as being by leaders in their particular fields, have that key virtue – a crucial one in a time of rapid change -of being (reasonably) up to date.

There are two underlying questions which we want to ask, both of which employ spatial metaphors for time. First, what is the legacy of the 1980s for the 1990s in housing policy? This is rather an artificial question, brought on by our having passed a marker in the flow of time to which we tend to attach great significance – the end of a decade. The fact is that if the early church leaders had been more accurate in their astrological computations the decade would have ended in 1994, or thereabouts; it is an artificial construct. Yet as a social construct it has power invested in it, the power to provoke a general feeling of retrospection and expectation. It also happens that the 'Thatcherite revolution' in British politics began around the beginning of the decade and petered out around its end, so the question has perhaps more logic than it ought to have.

The second question is directly related: is housing policy at some kind of crossroads? That it was at a crossroads in 1979 not many people would dispute, because we can look back and see a definite turning point in so many aspects of policy, notably the right to buy that was given to council tenants and the sharp decrease in expenditure on new building in the public rented sector. On the other hand, there were substantial continuities, such as the use of housing investment programmes to control expenditure by local authorities, the commitment to owner-occupation as the 'natural' form of tenure for those who could afford it, and the use of cuts in housing expenditure to achieve wider economic objectives. Was another turning point reached via the 1988 and 1989 legislation? Or are the continuities going to reassert themselves? And have we reached another crossroads in the 1992 election? And if we are at a crossroads, do we (or more particularly the opposition political parties) have much choice about which road to go down? These are the questions to which we will return in our conclusion.

The first part of the book is organised in a traditional way, with four chapters covering each of the major tenures (council housing, housing associations, home ownership and private renting) along with a chapter on one of the key institutions which sustain them – the building societies. The second part then provides a more explicitly value-based critique of current policy and practice, from three perspectives: race and gender, council tenants, and the 'new right'.

Peter Malpass (Chapter 1) provides a subtle treatment of the themes of continuity and change in relation to the local authority's role in housing. He begins with an analysis of the idea of the 'enabling' local authority, and argues that the combined effects of the 1988 and 1989 legislation will be to disable it. He takes a long-term historical perspective, from which it can be seen that an enabling framework *had* developed during the inter-war period, which lasted in a stable form until the early 1970s, encompassing the provision of both council housing and substantial support for private-sector housing. For him, the crossroads was reached in the mid-1970s, though it was in the 1980s that a 'new era' was entered, with a concerted attack by central government on both the providing and the enabling role, which intensified during the decade to reach 'new heights' of hostility to council housing.

He sees two legislative 'onslaughts' in the early and late 1980s, with the mid-decade being 'a phase of implementation or drift, depending on your point of view'. The second wave of legislation has been designed partly to remedy defects in implementation of the first, and its effects will not be felt until the measures taken in the 1989 Act have had time to take effect. His detailed explanation of this Act shows graphically that the problem of implementation is as crucial in housing policy as it is anywhere else. We might expect the emphasis to be on change, as in the breaching of the established principle of housing finance that general subsidy and means-tested assistance should not be rolled together. Yet Malpass shows how in some areas, notably the funding of council housing, there is an underlying continuity in aims underlying the whole decade, and that in general there is an underlying continuity in the residualisation of council housing traceable back to the mid-1950s.

It is fitting that Chapter 2, Mike Langstaff's chapter 'Housing associations: a move to centre stage', should be placed next because, as he explains, associations have been earmarked by the Conservative government as the major providers of new subsidised housing for rent; the alternative to council housing. The time frame is different; for associations the decade of change stretches back another six years, to 1974, when they were first given secure funding and an expanding role. The perspective on change is also different; despite an 'abrupt end' to expansion in 1980 – and other changes such as the right to buy for tenants of non-charitable associations and an increase in the delegated powers of the Housing Corporation which oversees associations – his conclusion is that with hindsight these changes are not now seen as so important. Stability rather than change was the outcome, and associations benefited from a period of 'benign neglect'. He explains that real change only came late in the decade, when the 1988 Housing Act fundamentally altered the public/private balance in the nature of the housing association by introducing a system of mixed public/private funding and a highly controversial new form of private tenancy which would also apply to new association tenants.

Langstaff's analysis of subsequent changes shows how even powerful Conservative governments have to ensure compliance. Even though some changes were made against the movement's wishes, associations were able to mitigate the effects of others, at least for a time. Yet, as with the impact of the 1989 Act on council tenants, the impact of the 1988 Act changes on association tenants will, it is feared, eventually be felt during the 1990s. Interestingly, while some changes are expected to be in the same direction – all tenants facing higher rents, for instance – others will be in opposite directions; while council housing may be broken up, association stocks may consolidate into larger and more powerful agencies.

The new emphasis on private funding for social housing is in some respects a very old one, in that associations in the last century relied on raising private loans and share capital. The new regime operates in a much more sophisticated capital market, in which lending is mediated through exchange institutions, the attitude of which is crucial to any understanding of the potential for future social housing developments. In Chapter 3, entitled 'Building societies: builders or financiers?', Douglas Smallwood analyses the way in which building societies have created, as well as reacted to, changes during the 1980s. Societies had already begun to plan for the eventual levelling off in the growth of owner-occupation; they foresaw a need to diversify into new markets, which, inevitably, meant some form of rented housing. Already, after the 1980 Housing Act allowed them to take part in a range of low-cost home-ownership initiatives, they had begun to work more closely with housing associations. Then the

Building Societies Act 1986, with its deregulation of the market, made it even more imperative that societies find new 'special housing markets'. Smallwood analyses the prefigurative forms of partnership which have influenced the late 1980s expansion, and shows how societies draw back from becoming developers or managers of housing, diversifying, but mainly within their own established field as lenders.

The chill winds of high interest rates and mortgage repossessions have, in the early 1990s, led to greater caution about helping government policy along when it involves operating at the margins. Yet there has been solid expansion into lending for associations and for voluntary transfers of council stock. Smallwood explains the limitations of private funding, and why low-start mortgages can only be a small part of the societies' lending and are no panacea for inadequate public subsidy. The prognosis is that as the mainstream mortgage market gets tougher, social housing business may be sidelined by societies fighting for market share among their 'existing customer base'. Yet some will continue to specialise in social housing, and will be needed, not only because of the sheer scale of the task ahead in providing new social housing but also for their expertise in co-ordinating partnership development programmes.

Stuart Lowe's Chapter 4, entitled 'The social and economic consequences of the growth of home ownership', enters into a controversial debate which has been raging among housing analysts throughout the 1980s concerning the significance of owner-occupation. Is home ownership an independent source of wealth and, if so, does this cut across the traditional predictor of life chances, social class? The underlying theoretical controversy, between Marxists and Weberians who set up different definitions of class, has reached a point where it can only be taken forward through empirical evidence and a sense of the changing pattern of inequalities over time. These Lowe provides for housing inheritance and equity leakage. He picks his way carefully through the debate, showing that owner-occupation 'has become for millions of households a significant generator of cash and wealth', which has indeed become constituted independently of class structure but which, because it is not a homogeneous tenure, 'does not bestow its benefits evenly'.

The relative autonomy of tenure grew during the 1980s, and its trajectory is increasing. But the effects vary by age, class and region, and over time in a 'house price cycle'. Gains are made at specific points in time, the obvious points being when people move or inherit a house. Less obviously, some people move but do not buy again because they marry, or enter residential care, and others release equity by remortgaging. The effects also vary over the life-cycle of the household, whose housing costs and benefits change drastically from youth to middle age, enabling older people to meet their welfare needs even when the public sector is declining. The conclusion is that an important consumption-based cleavage has developed between owners and renters which, combined with the residualisation that Malpass describes, means that the 1980s fuelled a potential dynamic for inequality which will be realised during the next decade.

In Chapter 5, 'Private rented housing and the impact of deregulation', A.D.H. Crook takes up another important issue: the government's determination to revive the private rented sector through deregulation. This was a late development of the 1980s, since the first two governmental terms were essentially taken up in promoting owner-occupation. Their success in this aim enabled them to switch attention to a small (and shrinking) tenure form which had been considered something of a backwater of housing policy analysis. Crook shows how the steepness of the decline during the 1980s has prompted a wide interest in the revival of the private sector, particularly for people who need to be mobile and gain ready access to housing. He identifies low demand and supply subsidies to home-owners, rather than rent and security legislation, as the main obstacle to reviving the sector, and illustrates this with a description of how each sub-sector of the market actually works; as with Lowe's analysis, the more the real world is analysed, the less the simplistic global statements that have characterised the debate hold up.

The 'central dilemma' is that, since most demand is from low-income households, and high rents must be charged to guarantee a reasonable return, subsidies are needed if the sector is to compete with the other, already subsidised, sectors. On the other hand, housing benefits which subsidise the tenant undermine hopes of a genuinely free market emerging. Another dilemma is that increased rents will propel those who can afford it into owner-occupation, thus reducing demand. By outlining the variety of measures which will have an impact – the Business Expansion Scheme, the new discretionary improvement grant system, and the likely increase in council housing rents – and by elaborating on both the types of tenant and types of landlord involved, Crook produces a very sophisticated analysis of the interaction of supply and demand within the new regime. He shows that, while some of the conditions needed for a revival are there, others – notably long-term political stability – are not. As with other recent policy changes, it will be the mid-1990s before we know whether this one has been successful.

Chapter 6, entitled 'The 1987 housing policy – an enduring reform?', is by a 'new right' academic, D.A. Coleman, who provides an interesting critique of government housing policy which takes for granted the latter's values and policy aims but questions whether the policies put in place by the 1988 and 1989 legislation are radical enough, whether they will actually achieve their aims of reviving the private rented sector, boosting housing association production through use of private funding, and transferring council housing to private landlords. His aim is 'to consider the reasons which persuaded the government to make such radical proposals in such a potentially politically dangerous area and to see how far the Housing Acts and related measures seem likely to realise their aims, within their own terms of reference'.

Coleman's analysis of the obstacles includes conventional ones such as tax relief to owners, non-market rents for social housing and the disadvantageous fiscal environment for private landlords; it also includes obstacles seen by the 'new right' in particular as being important: the effects of planning regulations on new building and the lack of a plausible private-sector alternative to housing associations for proposals to devolve local authority housing. So his prescription is equally unconventional: grants to private-sector landlords, both to counter the effects of mortgage tax relief and the effects of below market-level rents in the public sector. The Business Expansion Scheme is crucial here, and it is interesting to compare Coleman's analysis with that of Crook, in the previous chapter; his more optimistic prognosis is that it has begun the revival, but that more permanent incentives are needed, such as an English equivalent of the Welsh and Scottish agencies which are able directly to subsidise private landlords. He accepts that the current system of subsidies to both housing and people will continue for some time, even though he finds it confusing and argues for eventual transition to market rents coupled with housing benefits.

It is when Coleman comes to the transfer of local authority estates that perhaps the most radical part of his argument is revealed. Housing associations are regarded as a 'privileged', quasi-public sector, which are in some ways 'more like an extension of council housing, not a suitable replacement for it' and which, if they are to grow large enough to take over council housing, should be encouraged to privatise themselves.

In Chapter 7, 'Issues of race and gender facing housing policy', Norman Ginsburg and Sophie Watson offer an equally radical perspective on current policy, but from a standpoint totally different from those of both the 'new right' and the traditional 'Fabian' left. They given an account of the housing situation of ethnic minorities and women, concentrating on a statistical picture which shows that on average people in these groups are in worse housing than others. Ginsburg begins with a statistical description in relation to black people (including here other groups such as Asians). Making the most of inadequate data (lack of ethnic monitoring being symptomatic of the problem), he describes physical housing conditions, the incidence of homelessness, treatment of homeless people by local authorities, institutional racism in allocations practices, and what he calls 'structural racism'. This last concept includes the unintended or differential consequences of government policies, such as the cutbacks in spending on council housing, the impact of the right to buy, and support for home ownership.

Some of these disadvantages can be explained in the traditional way by reference to socio-economic class, but Ginsburg shows how this explanation is inadequate; active racial harassment and the threat of it are also causal, and so policies must be pursued to combat them at the local level. His prognosis for the effects of the recent housing legislation is gloomy; conditions in both the public and private rented sectors will get worse, and solutions, such as black housing associations and co-ops, will find it harder to provide new housing. Yet black people are not passive victims of housing policies; as resistance to Housing Action Trusts has shown negatively, and Tenants' Choice may show more positively, organised action may still have an effect.

Sophie Watson begins with a 'tenurial' approach, showing that access to owner-occupation is more difficult for women than for men. Given a background of changing patterns of family make-up, and the increasing participation of women in the workforce, she sees a policy to expand home ownership as 'the privileging of one household over another, and indirectly as privileging men over women'. Her prognosis for the public and private rented sectors is equally gloomy; since women rely on them disproportionately, and since conditions in those sectors will worsen as a result of government policy, the latter is unfair to women. Like Ginsburg, Watson sees action against Housing Action Trusts as signalling hope, but in this context because women are leaders in the struggle. Switching from a tenurial approach to one identifying categories of women in need – homeless, older women, elderly and disabled, divorced and separated, those suffering from domestic violence – she both weighs up the outcomes of current policies and suggests ways of strengthening such women's access to decent housing. Her perspective shows that new insights can be gained into other, previously undefined areas of research, such as the situation of black women and the lack of women's influence on housing design.

Finally, in Chapter 8, Johnston Birchall takes a consumer perspective on council housing, entitled 'Council tenants: sovereign consumers or pawns in the game?'. Like the last two chapters, this runs at a tangent to the mainstream of housing policy analysis, because it emphasises a particular set of values which are not normally accorded priority. It begins with the historical insight that council tenants were systematically excluded from exercising control of their housing alongside the other interests of housing workers and elected members. Within a strong framework of tenants' rights, it then compares the rhetoric of consultation, participation and so on, against the actual practice of council housing during the 1980s.

Because the focus is on the formal aspects of the landlord-tenant relationship – security of tenure, the tenancy agreement, information and consultation procedures, and strategies for collective empowerment of tenants – it could be argued that the critique is over-pessimistic, since human beings have the capacity, in their day-to-day interactions, to transcend the structures within which they have to work. Yet evidence drawn from some key studies undertaken during the 1980s shows that prior to the introduction of the 'Tenants Charter' in the 1980 Housing Act, the landlords resisted almost all the proposed tenants' rights, that during the early part of the decade many authorities tried to evade their new duties, and that even late into the decade progress in tenant consultation procedures was still quite slow.

Lack of space precluded a similar analysis of housing associations and the extension of the 'rights' approach into those rights relating to delivery of housing services (both of which are developed by Birchall elsewhere). However, his conclusion is quite clear: that in some important respects council housing is a prisoner of its own historical structures and vested interests. It cannot be presumed by those wishing to preserve affordable housing for low-income people that it occupies the moral high ground as of right. On the other hand, he goes on to analyse current central government policies such as Tenants' Choice, Housing Action Trusts and the transfer of new town housing, and local government policies such as voluntary transfers, and finds that the rhetoric of consumer choice is not matched by genuine commitment to offer tenants a full range of choices.

1 Housing policy and the disabling of local authorities

Peter Malpass

The White Paper on housing, published in September 1987, set out a series of policy objectives which included the following:

> The Government will encourage local authorities to change and develop their housing role. Provision of housing by local authorities as landlords should gradually be diminished, and alternative forms of tenure and tenant choice should increase. Some authorities will want to move in this direction themselves, and the Government will assist them. Some tenants will want to take the initiative and the Government will give them new rights to do so because this will enable them to improve their housing conditions and to have a say in their own future. *Local authorities should increasingly see themselves as enablers who ensure that everyone in their area is adequately housed: but not necessarily by them.*
>
> (DoE, 1987, p. 3) [emphasis added]

The idea that local housing authorities should become enablers rather than providers of housing has been much discussed and has taken root in some circles as a basic tenet of housing strategy for the 1990s. However, the two main themes of this chapter are that the legislation of 1988 and 1989 represented a negation of the enabling objective, and that as such it continued an established trend in housing policy which amounts to the disabling of local authorities. To some extent these themes take issue with the idea that housing policy is at a crossroads. The crossroads metaphor implies that choices have to be made about future directions, but with respect to council housing and local authorities' role in housing it is argued here that their route has been fairly clearly mapped out for some years past. And, to mix metaphors, this aspect of housing policy is rather like a supertanker: once embarked on a particular course it cannot quickly slow down or negotiate sharp changes of direction. The discussion is located in two related explanatory perspectives. First there is the notion of the restructuring of housing tenure, leading to the residualisation of council housing (Forrest and Murie, 1988; Malpass, 1990; Malpass and Murie, 1990). Second there is the wider context of central government attempts to manage the economy as a whole, and public expenditure in particular, which has led to tighter controls on local government.

The chapter is in three main sections. The first examines the extent to which in the past both central and local government played enabling roles. The middle section looks at policy developments in the 1980s, suggesting that a new era for council housing (Murie, 1982) began in the early part of that decade. The third section concentrates on the introduction of the new financial regime for local authority capital expenditure and for housing rents and subsidies in the 1990s.

THE ENABLING FRAMEWORK

There are two distinct aspects to the discussion under this heading. Central government policy on housing developed from its nineteenth-century origins through to the 1970s on the basis that the role of the centre was to enable local councils to respond in appropriate ways to local housing problems. This meant that local housing authorities operated in a policy environment characterised by Griffith in 1966 as *laissez-faire,* in the sense that the Ministry of Housing and Local Government left local authorities to decide what was their local need and how far, if at all, it was to be met by direct council provision (Griffith, 1966, p. 519; see also Wilks, 1987). Thus it is reasonable to emphasise the enabling role of central government, promoting and facilitating action by local authorities. In this context models of central-local relations which refer to partnership are appropriate (Rhodes, 1981; Houlihan, 1988, ch. 2).

A series of Acts in the second half of the nineteenth century gradually established and clarified local authorities' powers to build houses (Gauldie, 1974; Burnett, 1986; Hughes *et al.*, 1991). However, there was no compulsion, and until 1919 no financial support or encouragement from the Exchequer. The twenty years from the middle of the First World War to 1935 can be seen as the formulative period for council housing (Malpass, 1991). It was during that time that the basic enabling framework was established, followed by a long spell of relative stability until the early 1970s. The enabling framework was based on five key provisions: (i) local council control of capital expenditure programmes; (ii) Exchequer subsidies paid as fixed cash sums per dwelling per year for specified periods; (iii) rents set at 'reasonable' levels – in effect the levels determined by local authorities themselves; (iv) rent rebate schemes also subject to local determination; (v) the housing revenue account subject to a 'no profit' rule – surplus income could not be freely transferred into the general rate fund. It is important to note here that local authorities enjoyed wide discretion to make voluntary transfers in the opposite direction, from the general rate fund into the housing revenue account.

A feature of this financial structure was that it combined considerable local autonomy with limited financial liability for the Exchequer and a capacity for central government to steer or influence policy implementation at the local level. For instance, central government was able to affect the rate of new building by raising or lowering the subsidy level (Merrett, 1979, pp. 43–6), and to direct local authorities to general needs provision or slum clearance activity by attaching conditions to subsidies.

In addition to widely drawn local autonomy in relation to new building, rents and rebates, councils were free to determine their own approaches to issues of allocation and management of their housing stocks (Power, 1987). Indeed, in view of recent developments in the voluntary transfer of council housing to non-municipal ownership, it is interesting to note that under the Housing Act 1935, authorities could transfer all their housing functions to housing associations.

Turning to the enabling role of local authorities themselves, it is clear that from the outset councils possessed powers to assist private owners. Mortgage lending by local authorities, for instance, dates from the Small Dwellings Acquisition Act 1899. Holmans (1987, p. 230) reports that in the 1920s such lending was equivalent to between a quarter and a third of building society lending. In the post-Second World War period local authority lending was encouraged by legislation in the late 1950s (Smith, 1988, pp. 96–7). Home owners and landlords have also benefited from the local authorities' role as enablers of private-sector renewal and improvement. The provision of improvement grants dates from 1949. Housing associations, too, have been supported by local authorities since the 1930s, and for many years council loans were a main source of development finance.

It is true that these enabling activities were generally overshadowed by the scale of local authorities' own building programmes, but council housing itself should not be seen as entirely separate from the private sector. The important point here is that the construction of council housing was almost always carried out by private contractors, and in this sense local authorities acted as developers supporting the construction industry by generating profitable business. At certain times, particularly after the two world wars, local authority investment helped the private sector through very difficult periods. To this extent council housing was enabling of profitability in the private sector rather than presenting a challenge to it.

The role of local housing authorities as providers and enablers began to be seriously challenged from the late 1960s onwards, as a result of developments in the economy as a whole and the housing market in particular. A number of key housing policy developments, especially since 1970, can be understood in this context. The economic crisis which led to the devaluation of sterling in 1967 was followed by cuts in public expenditure and a steep decline in council house building. Then in the 1970s gathering economic difficulties and rising inflation produced policy responses which emphasised control of public expenditure. A revival of housing investment in the mid-1970s was halted by the most severe economic crisis since the war, and local authority housing capital spending fell rapidly in the second half of the decade (Malpass, 1990, pp 129–31). The situation at that time has been well summarised by Holmans in a reference to cuts in council mortgage lending: 'When public, political and financial attention was focused on the total of public expenditure and then in the 1970s the public sector borrowing requirement, one cut was as good as another, £ for £, and cuts in local authorities' house purchase lending were among the easiest to make' (1987, p. 254).

Attempts to reduce inflation led to the introduction of cash limits in 1975–76, and in housing the Labour government introduced in 1977–78 a system of limits on borrowing for capital expenditure in the form of Housing Investment Programmes (HIPs). The system meant that for the first time each local authority would have a planned ceiling for housing investment, a development which was justified in terms of both greater local freedom to spend within the agreed total and the need for overall control of public expenditure (DoE, 1977, p. 77).

At the same time as economic pressures were squeezing local authorities' role in housing, developments in the housing market were also pointing towards a more restricted role for council housing itself. The combination of social, political and economic circumstances which had underpinned the growth of a broad-based public sector in the aftermath of the war had been replaced by the 1970s by pressures to confine council housing to a more residual role. On the one hand, the continued growth of owner-occupation meant cutting back on council housing and encouraging the better-off council tenants and potential tenants to buy instead. On the other hand, the long-term decline of private renting meant that the least well off had increasingly to turn to council housing, and policies were required to make this sector more accessible to them.

The Housing Finance Act 1972 represented an attempt to residualise council housing and to reduce public expenditure on housing subsidies by introducing an approach to rents and subsidies which would make council renting simultaneously more expensive for the better off and cheaper for the less well off. The system of fair rents, deficit subsidy and mandatory rent rebates broke the mould of council housing finance which had endured since the 1930s, but in doing so it also removed local autonomy over rents and rebates. The reduction in local autonomy was a key factor in the early repeal of the fair rents provisions (Malpass, 1990, ch. 6), but nevertheless the Act did open the way for further bouts of reform in 1980 and 1989.

THE 1980s: A NEW ERA FOR LOCAL HOUSING AUTHORITIES

The 1980s was a decade when right-wing Conservative governments implemented economic and housing policies which combined to establish a new era for local housing authorities, an era in which their role as providers was greatly diminished and their capacity to act as enablers was weakened. Local government in general was seriously affected by central government economic policies which identified the level of public expenditure as being at the heart of Britain's problems.

Within the overall squeeze on public expenditure, local authorities suffered more than central government programmes, and there were numerous government attempts to control council spending, culminating in the poll tax (Burgess and Travers, 1980; Newton and Karan, 1985; Stoker, 1988).

Housing was the programme area singled out for cuts, and housing policy in the 1980s was consistently led by taxation and public expenditure considerations. One of the most striking features of housing policy in this period was the severity of cuts in public expenditure (although not in tax relief for mortgaged home-owners). In 1980 the government announced expenditure plans in which at least 75 per cent of all planned reductions were concentrated in the housing programme (House of Commons, 1980, p. v). And as O'Higgins (1983) noted, 'housing is not only the welfare programme suffering the largest cuts, it is also the only programme where cuts have been over-achieved.' The HIP system was used to squeeze local authority spending; expressed in 1986–87 prices, total HIP allocations in 1978–79 stood at £4,849 million, but by 1986–87 had fallen to £1,412 million (Malpass and Murie, 1990, p. 93). The HIP system had 'evolved into a narrower mechanism for short term financial control and the imposition of national housing policy objectives at local level' (Leather, 1983). However, as the decade progressed, so HIP allocations became less significant in the overall total of local capital expenditure, as a result of the growth of capital receipts from the sale of houses and land. By 1986–87 HIPs accounted for only 47 per cent of permission to borrow.

The cuts in capital expenditure were part of a three-pronged attack on council housing in the early 1980s. The other two parts were contained in the Housing Act 1980. Both involved significant reductions in local autonomy, and both were clearly designed to reduce and residualise public housing. The sale of council houses was promoted by the introduction of a statutory right for secure tenants to buy their houses or flats at substantial discounts from the market price (Forrest and Murie, 1988). This inevitably attracted greatest interest from amongst the better-off tenants living in the more desirable houses, but to the incentive of the discounted price was added the further stimulus of substantial rent increases for people choosing to remain as tenants. The 1980 Act introduced a new subsidy system which enabled central government to exert very powerful pressure on councils to raise their rents (Malpass and Murie, 1990, ch. 5; Malpass 1990, ch. 7). Taking the first Thatcher administration as a whole, average unrebated council rents in England and Wales rose by 119 per cent, during a period when the retail price index rose by only 55 per cent. This undoubtedly helped to fuel the demand from tenants wishing to buy their homes. Sales rose rapidly in 1980, topped 100,000 in 1981, and passed 200,000 in 1982.

Meanwhile, the squeeze on investment was reflected in falling levels of new building. In every year since 1919 until 1980 new building exceeded sales; in every year since 1980 sales have exceeded new building.

The production of new council dwellings fell in each year of the 1980s, with levels as low as 6,000 planned for the early 1990s (Treasury, 1989, table 9.8). The year 1980 represents a major turning point in the development of council housing to the extent that it marked the end of sixty years of growth and began a period of significant decline. As a proportion of all housing in Britain, the council sector stood at 31.5 per cent in 1979, and 23.6 per cent in 1989.

As providers of housing, therefore, local authorities were severely disabled by the measures introduced in the early 1980s. To the three elements of capital cuts, rent increases and the right to buy can be added a fourth disabling factor, namely a campaign of denigration of local authorities as inefficient housing managers. Council housing was transformed from being a solution into a problem. Housing ministers such as Patten and Waldegrave in 1987 and 1988 emphasised the failings of councils as providers of housing. Their criticisms were reinforced by two reports from the Audit Commission (1986a and 1986b), one referring openly to a crisis in housing management and the other emphasising problems in housing maintenance. To this was added a certain amount of influential academic criticism of council housing (Coleman, 1985; Power, 1987; Minford *et al.*, 1987). All this amounted to an assault on the confidence and credibility of local authorities as housing providers, especially when set in the context of the simultaneous lauding of home ownership in repeated official statements.

The reduction in local authorities' role as providers was not matched by an increase in their ability to act as enablers. To a large extent the enabling role is dependent upon financial resources and the discretion to dispose of them in the most appropriate ways to meet local needs. Constraints on finance in the 1980s curtailed local enabling activity in relation to the private sector. Mortgage lending, for instance, became negative after 1982, despite continuing increases in house prices (Smith, 1989, p. 180). The importance of finance is also well illustrated by fluctuations in renovation grant expenditure. In 1982–84 the government encouraged authorities to spend freely on grant aid, but when the funds were reduced by the centre the number of grants awarded fell quickly, by some 50 per cent between 1984 and 1987. Local authority aid to housing associations has also declined.

After its initial legislative onslaught on the housing role of local authorities, the government moved into a phase of implementation or drift, depending on your point of view, in the mid-1980s, but gathered itself for further radical legislation in 1988 and 1989. The next section of this chapter looks at the 1989 Act in some detail, but a brief reference to the Housing Act 1988 is relevant first. Whereas the policy emphasis of the early 1980s was clearly on the promotion of home ownership, by the latter part of the decade attention had turned to the continuing need for rented housing. The main thrust of policy contained in the 1988 Act was to deregulate the private rented sector and to establish a new financial regime for housing associations (Malpass and Murie, 1990, ch. 5). In addition the Act introduced two radical measures designed to further reduce local authorities as providers of housing; ministers had apparently convinced themselves that significant numbers of council tenants were so frustrated that they would willingly opt for a change of landlord. The Act therefore introduced a right for approved landlords to buy parts of the council stock, subject to a ballot amongst tenants (a ballot which was widely interpreted as rigged in favour of a positive outcome because abstentions counted as votes in favour of a change of landlord). In addition, the government proposed Housing Action Trusts (HATs) to be set up along the lines of urban development corporations and to take over sections of council housing designated by the Secretary of State to be in such great need of renovation as to be beyond the capacity of the local authorities.

Both the change-of-landlord and HATs proposals were clearly grounded in a highly critical and negative view of local government and represented a further wave of privatisation measures. In practice, however, their early impact was not what the government anticipated. Local authorities responded by adopting customer-oriented policies designed to improve housing management services, and for their part tenants showed both a reluctance to opt for different landlords and a marked hostility to HATs.

However, the full long-term impact of the 1988 Act may not become apparent until the measures in the Local Government and Housing Act 1989 have had time to take effect.

THE NEW FINANCIAL REGIME FOR LOCAL AUTHORITIES

This section looks at the new systems for regulating local authority capital expenditure and housing rents and subsidies. As the preceding analysis has argued, local authorities historically enjoyed considerable autonomy in relation to capital spending, rent setting and the level of rate-borne housing subsidy. Since the early 1970s successive governments have introduced measures to tighten control of local policies, but the local authorities have exploited opportunities to get round the restrictions imposed on them. The 1989 Act thus represents another attempt to reduce local autonomy in the context of a continuing central government concern with both the control of public expenditure and the residualisation of council housing.

Taking the capital side first, the legislative framework which operated in the 1980s was set up by the Local Government, Planning and Land Act 1980. From the point of view of both central and local government the 1980 system did not work satisfactorily, and in 1986 the government published proposals for change in a Green Paper, *Paying for Local Government*. Reaction to these proposals led to further work, and a consultation paper was published in July 1988 (DoE, 1988a). In the present context of discussion focused on housing, it is important to note that the proposals for reform of local authority capital expenditure apply to all authorities (including the county councils) and to all services, although housing accounts for the largest part of both outstanding debt and capital receipts from asset sales. By the end of 1986–87 local authorities in England and Wales had a total outstanding debt of £45 billion. This large sum is in fact rather modest when compared with the value of the assets owned by local authorities; the local authority housing stock in England and Wales was officially valued at £107 billion in 1989, and to this must be added the value of all other land and buildings owned by local authorities.

Nevertheless, central government maintains the view that local authority borrowing and capital expenditure must be controlled, for the following reasons:

1 The need to manage the national economy in view of the effect of local authority decisions on public spending and the public-sector borrowing requirement as a whole;
2 The need to ensure that investment by local authorities responds to national priorities;

3 To maintain accountability, since the financial effect of expenditure financed by borrowing is felt only to a very limited extent when it is incurred;
4 To safeguard the interests of future local taxpayers; and
5 To maintain the high credit standing which local authorities generally enjoy (DoE, 1988a, p. 7).

The 1980 system, however, was not able to meet the government's objectives, and was seen to suffer from four main problems. First, it failed to bring about capital expenditure at a local level which was consistent with public expenditure plans as a whole. Actual expenditure by English local authorities was wildly divergent from the government's annual cash limits in each year after 1981–82, with an underspend of 27 per cent in 1982–83 and an overspend of 52 per cent in 1985–86. The main factor in this situation was the difficulty in predicting both the growth of capital receipts (arising mainly from the sale of council houses) and local authorities' propensity to spend those receipts.

Second, the 1980 system created a distribution of capital spending power which did not match the need for expenditure. This arose from the fact that in distributing capital allocations through the HIP system, central government was not allowed (within the terms of its own legislation) to take into account the capital receipts available to individual local authorities. Capital receipts were regarded as an addition to spending determined by the HIP system, rather than a part of the total amount to be distributed. Unfortunately, capital receipts tended to be highest in areas with lower levels of need and lowest in areas with higher levels of need. In order to overcome this problem the government reduced the proportion of housing capital receipts that authorities could spend in any year, from an initial 50 per cent in 1980–83 to 40 per cent in 1984 and then to 20 per cent in 1985. The intention was to raise the proportion of resources within the total cash limit which could be distributed through the HIP mechanism. Another aspect of the problem posed by capital receipts arose from the drafting of the 1980 Act. The government's intention had been that only part (known as the 'prescribed proportion') of receipts would be available for new investment and that the rest would be used to pay off existing debts. In practice the government had to accept that the wording of the Act permitted authorities to spend not only the prescribed proportion of the current year's receipts but also the same proportion of accumulated receipts from previous years. This became known as the cascade effect and its impact added to the distortion of central government expenditure plans. It meant not only that local authorities could anticipate the gradual expenditure of all their capital receipts, but also that they retained spending power equivalent to the prescribed proportion of *notional* accumulated receipts, i.e. where the cash had been used to redeem old debt or to finance other permitted expenditure. It is important, in the context of the implications of the 1989 Act, to remember that authorities retained real flexibility in their use of the non-prescribed proportion of capital receipts. Some of the cash was used to pay off old debts, but some was used to finance capitalised repairs to the existing local authority stock, and this latter usage helped authorities to sustain capital programmes which were much larger than their HIP allocations.

The third main problem associated with the 1980 system was, from the point of view of central government, that it did not prevent local authorities from undertaking capital expenditure outside the framework of the legislation. During the 1980s local authorities proved to be adept at discovering and exploiting loopholes in the law in order to maintain levels of capital spending above those desired by the government. The 1989 Act was designed to close off any further exploitation of such loopholes. The fourth problem identified in the 1988 consultation paper was that because of the other problems discussed above there had been frequent changes in primary and secondary legislation, so that the 1980 system had not provided a stable framework within which long-term capital programmes could be efficiently managed.

In responding to these features of the 1980 system the government took into account four objectives:

1 To provide effective government influence over aggregate levels of local authority capital expenditure and borrowing;
2 To bring about a distribution of capital expenditure which reflects national and local needs;
3 To promote the government's aim of reducing the size of the public sector by asset sales and efficient asset management; and
4 To provide a sound basis for local authorities to plan their capital programmes with confidence (DoE, 1988a, p. 11).

The new system outlined in the 1988 consultation paper, and enacted in part IV of the Local Government and Housing Act 1989, is primarily concerned with controlling the use of credit by local authorities, as distinct from total expenditure, on the grounds that the poll tax will help to curb the proportion financed from revenue. The starting point of the new system is an attempt to close off any remaining loopholes by bringing together under one heading borrowing and all other credit arrangements which have the same economic effect as borrowing. There are two other sources of finance for capital projects:

1 Government grants or contributions from third parties (which might be other local authorities) and
2 Local authorities' own resources, including approved proportions of capital receipts and revenue contributions (in the case of housing projects, revenue contributions must come from HRA income).

Each year individual local authorities are given a 'credit approval' which places a limit on borrowing credit arrangements. This basic credit approval (BCA) may be supplemented by special credit approvals (SCA) issued by ministers to particular authorities in the light of circumstances during the year. In the case of housing, it is necessary to distinguish between two elements of the BCA: the HRA element and the non-HRA element. This arises because of the ring-fence around the HRA and the existence of two separate forms of central government assistance, the HRA subsidy and Revenue Support Grant. The HRA BCA represents the maximum volume of credit approvals on which the loan charges are eligible for subsidy, and the non-HRA BCA is used in calculating the capital financing element of Revenue Support Grant.

Under the new system the Exchequer provides capital grants (as distinct from annual contributions towards debt charges) in respect of certain local authority housing activities in relation to the private sector, including renovation grants, area improvement and slum clearance. There may also be situations in which an authority receives capital grants from another local authority. This might arise where a housing authority has obtained a large capital receipt from the sale of its entire housing stock and wishes to finance services in its area provided by the county council.

The third main element of capital finance for housing is capital receipts from the sale of assets. The new system permits the government to take into account an authority's own resources when calculating its credit approval level. The new system also provides for a proportion of capital receipts to be set aside for debt redemption or for the financing of future commitments. The proportion of housing receipts that local authorities can spend is 25 per cent, and 75 per cent must be used for debt redemption or future commitments. In the case of non-housing capital receipts, 50 per cent may be spent on new investment.

The new system effectively replaces the HIP mechanism, but the term HIP lives on in the context of the annual distribution of 'HIP allocations'. In November 1989 it was announced that the total HIP allocation for 1990–91 would be £1,789 million. This was the sum to be distributed by DoE regional offices using a combination of discretion and allocation according to the revised Generalised Needs Index. The HIP total was calculated from the sum of credit approvals (£1,346 million) plus a proportion of capital receipts (£270 million), plus specific capital grants (£468 million), minus £345 million kept in reserve for special programmes. Credit approvals plus capital receipts taken into account are referred to as Annual Capital Guidelines. In announcing the details of the package for 1990–91 the Secretary of State claimed that the new system enabled nearly two-thirds of resources to be allocated by the government in a way that took account of differences in local needs, compared with only one-third in recent years. The local authorities, however, have regarded the introduction of the new system with dismay because of the restrictions imposed on their spending and the implications for the HRA. The impact of the new capital controls varies from place to place, but it is clear that the intention is to give central government much tighter control over local authority expenditure. The application of capital receipts to the redemption of debt represents a severe restraint of capital programmes in many areas, but it also has the effect of reducing debt charges falling on the HRA. On the other hand, the elimination of the cascade effect places considerable pressure on rents to support continued expenditure on maintaining and refurbishing the existing council stock.

Turning now to the revenue side, in the early 1980s the government brought about a fundamental restructuring of local authority rents and subsidy policy, resulting in a major redistribution of assistance from general subsidy into rent rebates (housing benefit). In the short term this produced a sharp increase in the real level of rents, but it also had two other outcomes which became important reasons for further change. First, large numbers of authorities lost all subsidy and their HRAs moved into actual or notional surplus, thereby raising the issue of who was to control the size and use of such surpluses. For the years 1982–90 this question was effectively resolved by lobbying by the Association of District Councils in late 1981, the result of which was that the Secretary of State agreed to concede local control of surpluses. More than 25 per cent of authorities in England and Wales quickly acquired the habit of transferring HRA surpluses into the general rate fund.

The second relevant outcome of the 1980 subsidy system was that aggregate rate fund contributions to HRAs soon came to exceed Exchequer subsidy, something which had never happened before. Local authorities had become the major suppliers of 'indiscriminate' (i.e. non-means-tested) assistance, but in fact most authorities made either no rate fund contribution or just a very small payment.

During the 1980s London authorities generally accounted for around 75 per cent of total rate fund contributions, and since nearly all of the largest contributors were Labour-controlled authorities, this became a reason for the government's proposal, announced during the 1987 general election campaign, to ban all rate fund contributions.

It was in July 1988 that the government issued a consultation paper outlining its proposals for a new financial regime covering local authority rents and subsidies (DoE, 1988b). The consultation paper contained a critique of the 1980 system in which reference was made to the multiplicity of sources of subsidy (housing subsidy, rent rebate subsidy, rate support grant and rate fund contributions) and the diverse pattern of assistance across the country. The government's point here was essentially that it was not in full control of the flow of Exchequer resources into HRAs and that as a result of local decisions actual and notional HRAs were moving out of alignment. The second element in the critique was that the 1980 system had produced distortions in the incentives to efficiency and good management. It was argued that the freedom to make unconstrained rate fund contributions provided a cover for inefficiency in housing management. The same sort of argument was applied in reverse in relation to authorities generating surpluses in the HRA: 'It is essential that those surpluses should not be available to be used as a cushion for bad practices and inefficiency' (DoE, 1988b, para. 10). The stated objectives of the new regime were that it should be simpler, fairer and more effective. A simpler system, it was said, should produce subsidy arrangements which work in a more intelligible way and give consistent incentives. Fairness was referred to in relation to the balance between tenants and poll-tax payers, and between tenants in different areas. And an effective system would direct available resources to areas of need and provide an incentive for good management.

The new regime, effective from April 1990, is essentially a modified version of the 1980 system. Subsidy continues to be based on the notional deficit on the HRA in each local authority, and each year the Secretary of State issues determinations of the assumed changes in rent income and management and maintenance (M & M) expenditure. However, the new regime incorporates three important changes: i) the 'ring-fence' around the HRA, preventing contributions from the general fund (referred to as rate fund contributions before 1990) and discretionary transfers into the general fund; ii) a redefinition of the HRA deficit, and therefore what counts as subsidy; and iii) an attempt to differentiate increases in rents and M & M expenditure.

The ring-fencing provisions are straightforward, but require a brief explanation of the arrangements for cushioning the impact in those areas where large-scale transfers in either direction were the norm. In the case of authorities which previously subsidised housing from the rates, the new Exchequer subsidy effectively makes up the gap. The important point here is that central government now controls the whole of the subsidy required to bridge any HRA deficit and can therefore manipulate the deficit itself. In the case of authorities which previously transferred housing surpluses into the general rate fund, there is a transitional arrangement by which central government provides an annually decreasing amount of assistance to the general fund in compensation for the lost income.

The other provisions of the new regime are a little more complicated. Under the 1980 system authorities could qualify for deficit subsidy and rent rebate subsidy; it is important to understand the differences between them. Deficit subsidy, payable under the Housing Act 1980, made up the notional difference between total HRA expenditure and total income from *unrebated* rents (i.e. net rents actually charged to tenants plus rent rebate subsidy), plus any other income such as interest on capital receipts. By the mid-1980s most authorities were in notional surplus and so in those cases no deficit subsidy was payable, but rent rebate subsidy continued to be payable to all authorities, typically providing rather more than a third of all HRA income. The level of rebate subsidy was determined by the sum of the individual housing benefit entitlements of council tenants and not by the state of the HRA. Therefore, central government saw itself as contributing to HRA surpluses in a quarter of all local authorities, and paying subsidy in excess of what was needed in these areas.

The new regime deals with this problem by, in effect, moving to a new definition of HRA deficit based on *rebated* rents, and introducing a single deficit subsidy, known as the HRA subsidy. This new subsidy replaces the 1980 Act subsidy, rate fund contributions and rent rebate subsidy, and its purpose is to make up the notional deficit arising where HRA expenditure exceeds income from rebated rents plus other items. From the point of view of central government, this has two main advantages: it limits subsidy to the level required to balance the HRA, and it restores leverage over rent increases. Under the 1980 system leverage on rents could be exerted by reducing subsidy on the assumption that rents would rise faster than M & M expenditure. However, leverage was lost as authorities ran out of subsidy, but the new regime massively expands the definition of HRA deficit, and therefore correspondingly increases the amount of subsidy which can be influenced by the annual determination of changes in rents and M & M. Whereas in the late 1980s most authorities were not vulnerable to direct financial pressure on rents via the subsidy system, in the 1990s the power of that system has been restored and rent increases can be again determined by central government. The importance of this is that the amount of leverage which the government can exert on rents is closely related to the magnitude of the deficit, and without increasing public expenditure at all the new regime has expanded the definition of aggregate

HRA deficit from around £500 million to well over £3,000 million. However, in order to do this the government has had to roll together general subsidy and means-tested assistance, breaching an important principle in housing finance.

Turning to the differentiation of determinations for changes in rents and M & M, the underlying reason for a new approach was that standard determinations (as incurred in the 1980s) failed to take sufficient account of variations in local circumstances. The problem, however, is to find ways of differentiating the determinations accurately and fairly across the country. To achieve this goal requires the collection and analysis of large quantities of data, on a scale and with a degree of sophistication which has not been achieved in the past and is not readily achievable even now. The new regime was, therefore, introduced using the best available method for differentiating rent increases, and with no real differentiation of M & M increases.

In the case of rents, the 1988 consultation paper said only that 'rents should generally not exceed levels within reach of people in low paid employment, and in practice they will frequently be below market levels' (DoE, 1988b, para. 11). Ministers have been no more precise in subsequent statements, but they have said that they are not pursuing market rents in the public sector. It is clear, however, that the government wants rents to vary in a way which reflects the types of variations found in the private sector. The best available proxy for capital values is the valuations of council houses sold under the right-to-buy, and in June 1989 it was announced that differential determinations would be based on local right-to-buy valuations. The method employed by the DoE is based on a process which begins with the capital value of each authority's stock expressed as a fraction of the total value of all council dwellings in the country. If an authority's stock is worth, say, 1 per cent of the total stock then it would be expected to produce 1 per cent of the total rent income in the whole country. This gross amount divided by the number of dwellings would give an initial indication of the average increase (or decrease) for the year. The next stage involves a percentage increase reflecting the Secretary of State's view of how much rents generally should rise in the year, followed by a further amount to cover inflation.

When this approach was first tested, it was assumed that rents would rise by 5 per cent, plus 5 per cent for inflation. On the basis of capital values, however, a large number of authorities, mainly in the north of England, emerged from the exercise with substantial *reductions* in notional rents, reflecting their low capital values. For instance, Middlesbrough was shown to have a notional decrease of £15.88 per week, because it was already a relatively high rent authority in an area with low property values.

Elsewhere the data predicted increases of massive proportions, topped by £42.03 per week in the City of London.

Variations on this scale inevitably required a damping mechanism, and it was decided that in 1990–91 the calculation of subsidy would assume increases between 95p per week and £4.50, depending on the circumstances in each local authority. Over 92 per cent of authorities in the three northern regions of England were assumed to make the minimum increase, compared with none in London. Altogether 18 per cent of authorities were assumed to make the maximum increase, and of this group of 66 all but two were in London, the southeast or the eastern region. Thus rent increases were highest in areas which already have the highest rents, and the damping mechanism seems likely to become a permanent feature of the new regime.

On the issue of M & M expenditure, a major deficiency of the 1980 system was that the assumptions about expenditure were based on the updated figures which happened to apply in each authority in the late 1970s. By the end of the 1980s the M & M assumptions were poorly related to actual need to spend in different areas, and the government hoped to introduce the new regime with differential M & M allowances related to differences in the age and type of stock in each area, and to take account of geographical factors. However, this was not possible and in the first year of the new regime the basis of calculation was essentially a rolled-forward version of the old system. The most important points were, first, that the determination for 1990–91 was based on an increase in M & M expenditure of 5 per cent for inflation plus 3 per cent, which meant that expenditure was assumed to increase more slowly than rents; in other words, tenants could not expect to see the full value of higher rents reflected in better-quality services. Second, the government took no account of the fact that some authorities had been funding a high level of repairs expenditure from capital receipts, which were no longer so plentiful after April 1990.

CONCLUSION

In concluding this discussion there are a number of brief points to be made. It is clear that by the late 1980s government hostility to councils as providers of housing had reached new heights, representing a challenge to council housing which was more profound than previous fluctuations in the emphasis between private- and public-sector production. In this context, policy statements began to refer to authorities playing an enabling role, facilitating provision by non-municipal agencies. What has been argued above is that these developments have to be seen against a background of a long-term decline in council housing and a contemporary financial and legislative environment which is, in practice, disabling of local authorities.

The residualisation of council housing has been discussed for more than a decade, and its origins can be located further back in time than the election of the first Thatcher government. Indeed, it is possible to argue that the beginnings of residualisation were in the major policy changes in the mid-1950s. Other candidates to be identified as key crossroad years in the rise and fall of council housing would be 1968 and 1980. Whatever precise view is taken of these historical turning points, it is clear that at the start of the 1990s council housing is well on the road to a residualised condition. Debate no longer dwells on the growth of council housing, but focuses instead on issues of contraction. This view is reinforced by recognition of the extent to which developments in council housing are driven by the long-running restructuring of the private housing market and wider political and economic factors.

The enabling role of local authorities also needs to be understood in this context. Enabling requires resources, whether in money or land, and it requires that local authorities have the freedom to act imaginatively, flexibly and promptly. In reality, however, authorities are starved of financial resources, their reserves of land are declining and they are increasingly constrained by central government's highly interventionist approach. The 1989 Act develops the disabling of local authorities, first by reducing their ability to deliver high-quality services to their own tenants, and by constraining the supply of capital finance available for housing expenditure. The government even came very close to making it virtually impossible for authorities to participate in partnerships with private builders (Rees, 1990).

The 1989 Act represents a lack of trust in local authorities and contempt for their achievements in the past. However, whilst there is a continuing need for new rented housing there is no evidence that the private sector can meet this need at rents that people can afford, nor that the housing associations can rapidly and efficiently take over the dominant role in the provision of social rented housing. Indeed the evidence in 1989–90 suggests that the associations and the Housing Corporation have so far made a very confused and unsatisfactory start to their new role. Just as local authorities have been unfairly condemned for their past performance, too much faith has been placed in the housing association sector. What is required now is a much more even-handed approach, establishing a level playing field on which local authorities can compete on fair terms. This would mean, for instance, that councils would be permitted to determine their own capital programmes once again, borrowing against their asset base (the value of their existing houses). There can be no going back to past practices, and any view that councils should be given an enhanced role in housing should not be criticised on those grounds. The 1990s could ultimately prove to be a crossroads for housing authorities, opening up new opportunities to play a pivotal role in both direct provision and enabling services. But that prospect depends upon a change of government and a breadth of vision which is still to be developed.

REFERENCES

Audit Commission (1986a) *Managing the Crisis in Council Housing*, London: HMSO.
——(1986b) *Improving Council House Maintenance*, London: HMSO.
Burgess, T. and Travers, T. (1980) *Ten Billion Pounds: Whitehall's Takeover of the Town Halls*, London: Grant McIntyre.
Burnett, J. (1986) *A Social History of Housing 1815–1985*, 2nd edition, London: Methuen.
Coleman, A. (1985) *Utopia on Trial*, London: Hilary Shipman.
Department of the Environment (DoE) (1977) *Housing Policy: A Consultative Document*, London: HMSO.
——(1987) *Housing: The Government's Proposals*, Cmnd 214, London: HMSO.
——(1988a) *Capital Expenditure and Finance*, a consultation paper, July, London: DoE.
——(1988b) *New Financial Regime for Local Authority Housing in England and Wales*, a consultation paper, July, London: DoE.
Forrest, R. and Murie, A. (1988) *Selling the Welfare State*, London: Routledge.
Gauldie, E. (1974) *Cruel Habitations: A History of Working Class Housing 1780–1918*, London: Allen & Unwin.
Griffith, J.A.G. (1966) *Central Departments and Local Authorities*, London: Allen & Unwin.
Holmans, A.E. (1987) *Housing Policy in Britain: A History*, Beckenham: Croom Helm.
Houlihan, B. (1988) *Housing Policy and Central-Local Government Relations*, Aldershot: Avebury.
House of Commons (1980) *First Report of the Environment Committee, Session 1979–80*, HC 714, London: HMSO.
Leather, P. (1983) 'Housing (Dis?) Investment Programmes', *Policy and Politics*, vol. 11, no. 2, pp. 215–27.
Lowe, S. and Hughes, D. (1991) *A New Century of Social Housing*, Leicester: Leicester University Press.
Malpass, P. (1990) *Reshaping Housing Policy: Subsidies, Rents and Residualisation*, London: Routledge.
——(1991) 'The Financing of Public Housing' in S. Lowe and D. Hughes, *A New Century of Social Housing*, Leicester: Leicester University Press.
Malpass, P. and Murie, A. (1990) *Housing Policy and Practice*, 3rd edition, London: Macmillan.
Merrett, S. (1979) *State Housing in Britain*, London: Routledge & Kegan Paul.
Minford, P., Peel, M. and Ashton, P. (1987) *The Housing Morass*, London: Institute of Economic Affairs.
Murie, A. (1982) 'A New Era for Council Housing', in J. English (ed.) *The Future of Council Housing*, Beckenham: Croom Helm.

Newton, K. and Karan, T.J. (1985) *The Politics of Local Expenditure,* London: Macmillan.

O'Higgins, M. (1983) 'Rolling Back the Welfare State: The Rhetoric and Reality of Public Expenditure and Social Policy under the Conservative Government', in C. Jones and J. Stevenson, (eds), *The Yearbook of Social Policy in Britain 1982,* London: Routledge & Kegan Paul.

Power, A. (1987) *Property Before People,* London: Allen & Unwin.

Rees, A. (1990) 'Partners in Housing', *Housing,* May, pp. 24–9.

Rhodes, R. (1981) *Control and Power in Central-Local government Relations,* Farnborough: Gower.

Smith, M. (1988) *Guide to Housing,* 3rd edition, London: Housing Centre Trust.

Stoker, G. (1988) *The Politics of Local Government,* Basingstoke: Macmillan.

Treasury (1989) *The Government's Expenditure Plans 1989–90 to 1991–92,* Cmnd 609, London: HMSO.

Wilks, S. (1987) 'Administrative Culture and Policy Making in the Department of the Environment', *Public Policy and Administration,* vol. 2, no. 1, pp. 25–41.

2   Housing associations

A move to centre stage

Mike Langstaff

In the early 1990s housing associations will have an increasingly significant role in UK housing policy. The Conservative government has made its views clear. In the words of one of the ministers responsible for housing in the 1980s: 'Our new policies put housing associations on centre stage. We see them as the major providers of new subsidised housing for rent and as possible alternative landlords for dissatisfied council tenants' (Trippier, 1989). By 1990 associations had already overtaken local authorities as providers of new rented housing, and by 1992 their programme is set to double to 35,000 new and rehabilitated homes per year. In the same period council building will decline still further, to a level of 6,000 homes per year. The transfer of the existing council housing stock to housing associations is not proceeding as rapidly as the government envisaged, but it has been unofficially suggested that by 1996 the housing association sector will overtake the local authority sector through this process (NFHA, 1990c).

This chapter focuses almost entirely on changing central government policies and their implications for the 1990s. This is not because demographic changes, economic processes and technological development are unimportant, but because the changing shape of social housing in the UK is dominated by the impact of political initiatives (Back and Hamnett, 1985; Paris, 1989). It explores the impact of the Housing Act 1988 and associated measures on the work of housing associations. How and why did associations move to 'centre stage' in the government's thinking? What are the implications of the new financial regime for them and their tenants? What are the implications of growth and the transfer of risk to them? How might political events in the early 1990s affect the current direction of change? These are some of the questions addressed.

It should be acknowledged that the chapter does not provide a comprehensive description of the work of associations. Surprisingly, no book has tackled that task until recently: now the gap is filled (Cope, 1990). It should also be noted that the discussion and any statistics quoted in the chapter refer only to England, not to the UK as a whole. This is because there are some significant differences in the legislation and procedures affecting housing associations between England, Wales, Scotland and Northern Ireland, despite the general thrust of government policy being the same in each of these countries, and also because of the split of the Housing Corporation and Federations of Housing Associations into different national bodies.

EXPANSION AND BENIGN NEGLECT

The events leading up to the Housing Act 1988 and the prospects for housing associations in the 1990s can only be understood with some reference to the history of associations and, in particular, to the impact of the Housing Act 1974 in the 1970s and 1980s. In the fifteen years following that Act the housing stock of associations trebled in size to nearly 600,000 homes. Although this development activity was reduced in the 1980s, Thatcherite reforms avoided any fundamental change to associations' funding and direction until late in the decade.

Housing associations are diverse, non-profit-making organisations with a variety of constitutional formats, structures and aims (Cope, 1990, chs 1 and 2). They have different origins: a few can trace their history back to almshouses of the middle ages; some well-known associations developed as part of the nineteenth-century philanthropic housing movement; others were established by professional people with a vested interest in fee-earning (subsequently banned) during the brief flurry of activity in cost-rent and co-ownership housing in the 1960s; and still others grew as inner-city based, Shelter-backed organisations in the late 1960s, primarily undertaking rehabilitation work. Some 2,700 associations serve a variety of needs groups, are of varying sizes and cover varying geographical areas – from the very local to the national. Housing co-operatives are tenant-controlled associations (Birchall, 1988).

Whatever their origin or type, housing associations were eclipsed by the growth of council housing until the 1970s. The Housing Act 1974 then brought about a very favourable system for the growth of housing associations. Introduced by a new Labour government, it included much of what was proposed by the previous Conservative government. It considerably extended the role of the Housing Corporation, which had been established by the government ten years previously. The Housing Corporation was now to be the primary bank manager, watchdog and, to a lesser extent, advocate for associations' work. The Act also introduced the subsidy system which fuelled associations' expansion in the 1970s and 1980s. The major new subsidy introduced was Housing Association Grant (HAG) – a capital grant equal to the sum required by an association to reduce its loan repayments in the first year after completing a development scheme to the amount equal to its income from the fair rents set by the rent officer service (net of defined allowances for management and maintenance costs). In practice HAG wiped out, on average, 85 per cent of the costs of schemes and reached 100 per cent for many special needs schemes (Hills, 1987a). Moreover, this capital grant was buttressed by discretionary revenue grants (Revenue Deficit Grant (RDG) and Hostel Deficit Grant (HDG)) covering annual deficits which arose because of stock development under pre-1974 regimes, the effect of rent restrictions after the calculation of HAG and the high costs of managing hostels.

This generous subsidy system was accompanied by substantially increased funding for housing associations' activity via the Housing Corporation. By the late 1970s associations were receiving approvals for new schemes at the rate of about 50,000 homes per year. Initially, most effort went into new building activity, but rehabilitation work grew until it accounted for half of associations' production. Prior to the 1970s, virtually all rehabilitation work had been confined to London: now, as the government emphasised the contribution that associations could make to local authorities' housing renewal strategies, such work spread to most cities and large towns (Gibson and Langstaff, 1982). A high level of correlation between housing association rehabilitation progress and areas of housing stress was displayed (Kirby, 1981).

After the election of the Conservative government in 1979 this expansion came to an abrupt end. Halfway through 1980–81 a moratorium on further Housing Corporation approvals for new schemes during the remainder of the year was introduced: only 11,000 approvals were made in that year. Thereafter, the Housing Corporation-funded programme ran at about half the rate it reached in the late 1970s, with 15–20,000 new approvals each year until the Housing Act 1988. Notwithstanding this cut in the number of homes funded by the Corporation, its relative significance in funding associations increased. The 1974 Act had produced a dual system of funding by the Housing Corporation and individual local authorities. As local authorities' capacity to fund was restricted by drastic cuts in their Housing Investment Programme (HIP) allocations, the Corporation's share of reduced total spending on associations' development programmes increased. In 1977–78, 61 per cent of funding was channelled through the Housing Corporation; by 1982–83 it was 87 per cent.

There were other changes affecting the role of housing associations in the early 1980s. The Housing Act 1980 introduced the right to buy for tenants of non-charitable associations (the government having been defeated in the House of Lords when attempting to do the same for tenants of charitable associations): by 1989 some 19,000 homes had been sold. The same Act widened the powers of housing associations to carry out low-cost home-ownership schemes – shared ownership, improvement for sale, and leasehold for the elderly – and for a few years an increasing share of the Housing Corporation's programme was devoted to such schemes. By 1989, 36,000 homes had been provided under these arrangements.

The new Conservative administration also carried out a review of the role of the Housing Corporation in 1980 and gave it increased delegated powers, ending the 'double scrutiny' system under which the Department of the Environment also had to approve every housing association development. The Corporation was given the responsibility for managing its funding within an annual cash limit, with no tolerance for over-spending or under-spending. A block system of spending within its Approved Development Programme (ADP) was introduced, with Block 1 being expenditure on schemes already committed before the start of the year and Block 2 being expenditure on previously approved schemes starting on site. Block 3 expenditure on newly approved schemes represented the 'seed-corn' of new activity, and allocations to associations were made subject to an annual bidding process. Each association had to bid within a Housing Corporation region, whose share of the national programme had been determined by the use of an 'objective' statistical method of measuring relative housing need, the Housing Needs Index (HNI), itself a modified version of the system used for local authorities' HIP allocations (Langstaff, 1984).

Important as these changes seemed at the time, with the benefit of hindsight what is most notable about the early and mid-1980s is the stability afforded to housing associations' development programmes, as well as the persistence of the favourable subsidy system introduced in the 1970s. Much else was changing in housing policy under Thatcher, but housing associations went through a period of benign neglect.

First, although cut by half compared with the level of the late 1970s, housing association development was not savaged like that of local authorities. The changes affecting council housing are well documented elsewhere (see Chapter 1). What is relevant in this context is that housing associations were relatively favoured in capital spending and also that their work in focusing on inner-city rehabilitation and special needs provision remained constant. Furthermore, while local government was identified by central government as being at the heart of the country's economic and social problems – by allegedly spending too much, adopting 'loony left' plans, and being managerially ineffective – housing associations were still generally praised by Conservative spokespersons.

Second, the favourable subsidy system introduced by the 1974 Act remained intact. The Labour government's Green Paper on housing in 1977 had flirted with the abolition of HAG and its replacement with annual revenue subsidies, with local authorities being subject to the same system. It also proposed ending the fair rent system (Morton, 1989, p. 41). The Conservative government of 1979 also considered more fundamental change but in the Housing Act 1980 restricted itself to modifying the system by adding Grant Redemption Fund (GRF). GRF enabled the government to recoup the rental surplus which would normally arise when rents increased because residual HAG had been paid on the basis of conventional, fixed-interest loans. Income from the GRF was recycled to the Housing Corporation's ADP.

Just as important were the stable development procedures adopted by the Housing Corporation. Development activity was highly standardised, with the Corporation being a 'one-stop shop' for loan and HAG funding. The system protected all associations from commerial risks and allowed new associations to proceed with a relatively smooth learning curve when starting development activity. The block system of expenditure within the ADP was used by the Housing Corporation to ensure that it spent its cash limit exactly. Each association had to bid for Block 2 and Block 3 allocations and also secure specific project and tender approvals for each scheme, allowing the Corporation to control when development took place and when the variable part of its expenditure would be incurred. The system required a backlog of projects waiting to start on site and produced irritating 'stop-go' patterns of activity, but in acknowledgement of their lack of control over the pace of development, associations themselves were insulated from risk as capitalised interest was built into the final HAG calculation. Moreover, cost overruns were also met by HAG, provided that they were not caused by higher standards than those specified by the Corporation or by foreseeable items. In summary, the system was highly centralised and created a safe operating environment for associations.

RENTS, RISKS AND RIGHTS

This environment, however, was only enabling the production of new or improved homes by housing associations at half the rate achieved in the 1970s. During the mid-1980s there was a rapid growth of interest amongst housing practitioners and academics in mechanisms for attracting private investment into rented housing in order to increase production. This interest conformed to the government's general economic perspective that private, rather than public, financing of projects has a beneficial effect on general interest rates. Pilot schemes were put in place, and eventually the Housing Act 1988 introduced a system of mixed public/private funding, with associated changes to the rights and rent-setting procedures for new housing association tenants. The same Act introduced the Tenants' Choice legislation for council housing.

Some of the interest in private finance was directed at the possibility of reviving the role of the private landlord, but much was concerned with the funding of housing association development. In this latter area the perceived key to progress was using public grants to lever private loan finance in a way that would enable only the grant to count as public expenditure under the Treasury's spending conventions. In this way more homes could be produced for the same amount of public expenditure. The essence of the Treasury position was that private sector lending coupled with guarantees constitutes a contingent liability. Private finance must bear some risk, increase efficiency, or both (Kleinman, 1987; Foster, 1986). This implied a transfer of risk in development to associations and the end of fair rents, of RDG and the availability of HAG for future major repairs. A more confusing aspect of the Treasury's position was a concern about the maximum proportion of total scheme costs that could be covered by grant while still enabling the private contribution not to count as public spending. As we shall see, this changed over time.

In 1985 the report of a major independent inquiry into the future of housing policy was published (HRH The Duke of Edinburgh, 1985). One of its key recommendations was that rents in all sectors should in future be related to the capital value of properties to attract large-scale investment from pension funds, building societies and other institutions. It argued that a public funding monopoly made housing associations vulnerable. The proposal was part of a package of linked recommendations which also included a means-related housing allowance (with mortgage tax relief for owner-occupiers being phased out); an emphasis on the strategic role of local authorities; and new mechanisms for improving housing conditions.

Also in 1985 the Housing Corporation produced its first Corporate Plan, which argued for an increase in its programme to 40,000 units per year. In 1986 it was conceded in Wales that a privately funded scheme might attract up to 30 per cent HAG without the remaining costs being regarded as public spending when the development of 600 homes in Cardiff, Wales was approved. In the following year North Housing Association created considerable interest with its ambitious scheme to provide 3,000 rented homes, initially using purely private finance raised through stock issues, although also with the benefit of cheap land from local authorities and injections from its reserves. In 1987–88 the Housing Corporation launched a pilot programme of mixed public/private funding with 30 per cent HAG for 'challenge funding' schemes which associations considered they could develop on this basis and for shared housing for 'job movers'; and with 50 per cent HAG for temporary housing for homeless families. All these early pilot projects depended on the use of assured tenancies, which had been introduced by the Housing Act 1980 but little used by private landlords, so that initial and subsequent rents could be set by associations, not by the rent officer service. They also depended on private finance being made available on a low-start basis, with higher loan repayments in later years as rental income increases.

Moving on from its overwhelming emphasis on promoting owner-occupation and cutting local government expenditure, the Conservative government had already targeted rented housing for reform before the 1987 general election. After the election a White Paper set the scene for subsequent legislation. It stated that:

> The Government believes that housing associations have a vitally important part to play in the revival of the independent rented sector. During recent years they have been almost alone in the independent rented sector in investing in rented housing, and have played a key part in developing new styles of management and new forms of low cost home ownership. It will be important to build on this success and develop it further.
>
> (DoE, 1987a, p. 12)

A later document in that year put more flesh on the bones by explaining that the regime to be introduced in new legislation would have two aims. First, it would increase the volume of rented housing that associations could produce for any given level of public expenditure. Second, it would 'create new incentives to associations to deliver their service in the most cost-effective manner, bringing to bear the disciplines of the private sector and strengthening the machinery of public support' (DoE, 1987b, para. 2). Although there were a number of motivations behind the changes in the housing association funding regime (Kearns, 1988), these two reasons were dominant.

Debate before the Housing Act 1988 was passed and immediately after concentrated on the implications of the new regime in the areas of rents, risks and rights for associations and their tenants (NFHA, 1987b). The fair rent system introduced for private-sector tenants in 1965, and extended to housing association tenants in 1972, would no longer apply to new association tenants after the Act was passed. Existing secure tenants retained their status and the right to the registration of a fair rent by the rent officer service, but re-lets in the existing stock and lets of newly developed homes would be under the assured tenancy system with no rent control. The system was intended to ensure that rents for tenants of new schemes would have to meet the costs of loan repayments under a system of fixed (and lower) levels of HAG. Associations were to be expected to set rents for re-lets at similar levels as those for new schemes. Whereas in the previous system the rent had determined the grant applicable, the new system was intended to reverse this process. Rents would also rise because of the need for associations to establish sinking funds for future major repairs to their properties, which were previously assumed to be eligible for 100 per cent 'second bites at the cherry' capital grants.

The most important issue in this new system was the level of grant available. As we have seen, the pilot schemes of mixed funding started with 30 per cent HAG. A leaked draft of the DoE's consultation paper had referred to the natural and equitable limit of HAG being 50 per cent on average. This level of HAG would have produced enormous increases in the rents paid by new tenants, at a time when many asso-ciations were finding that the rents charged to existing tenants were becoming difficult to collect because of changes in the housing benefit system (CES Ltd, 1989). Critics argued that the changes in the funding regime would inevitably drive associations up-market in terms of whom they housed (Shelter, 1987; Hills, 1987b). The National Federation of Housing Associations (NFHA) tried to associate the issue of rents with that of rights by opposing the linking of new tenants' status with that of new private tenants under the assured tenancy regime, instead arguing for the creation of a separate housing association tenancy.

In the event, the key elements of the government's proposals were enacted, although new tenants were, in practice, given the same rights as existing ones by the Tenants' Guarantee, effectively a mandatory code of housing management practice enforced by the Housing Corporation (Housing Corporation, 1988). In that document backing was also given to promises made by ministers about rents during the passage of the Act by its confirmation that associations are 'expected to set and maintain their rents at levels which are within the reach of those in low paid employment' (Housing Corporation, 1988, para. D2), although further clarification of this statement was studiously avoided. Negotiations between the DoE, Housing Corporation and NFHA produced detailed development procedures which incorporated an average 75 per cent rate of HAG.

These detailed procedures achieved the government's purpose of transferring the risk in development to associations by grant levels being determined at the outset of schemes. Cost overruns would be paid for by higher loans and thus higher rents, not higher grants. A complex system of total cost indicators, incorporating notional on-costs for items such as consultants' fees, was combined with fixed grants to create incentives for associations to be more cost effective. While the basic mixed-funding procedures allowed some protection for associations if costs rose, experienced associations could elect to develop under a tariff system with fixed levels of HAG for each type of dwelling produced and accept the 'swings and roundabouts' involved. The use of private finance for loans on the best terms negotiable became the norm and, with a progressive movement of the Corporation's programme from public to mixed funding, the financial viability and sophistication of associations became more important. Although the new system was primarily about development risk, it also put a premium on curbing the costs of housing management at a time when good practice was being encouraged (NFHA, 1987a) and on planning for maintenance, including making provisions for future major repairs (Langstaff, 1988b). New schemes would not be subject to any clawback of grant as rents rose but a modified GRF was retained, now known as the Rent Surplus Fund (RSF). For schemes developed with old HAG, 70 per cent of surpluses have to be retained to build up major repair provisions; 15 per cent can be retained for other uses; and 15 per cent paid over to the government. The NFHA's aims in negotiations about these elements of the new regime were to restrict the impact of the changes on types of programme and associations threatened by, for instance, retaining public funding for special needs programmes and smaller associations, and to abolish RSF so as to allow associations to build up reserves as a bulwark against risk. In the former it was partially successful: in the latter it failed.

A further element of the Housing Act 1988 affecting housing associations was the Tenants' Choice legislation, giving council tenants the right to enforce a transfer of ownership of their homes to an alternative landlord (assuming such a landlord was willing to take on the role). The legislation is intended to follow on from the right to buy by creating more choice and reducing the role of council housing. Housing associations had already become involved in taking on the ownership of public housing from new towns under the Housing and Planning Act 1986 and, in the absence of a plausible private-sector alternative, were 'chosen to be chosen' by the government in this role (Langstaff, 1988a). The Housing Corporation was given considerably extended powers to promote Tenants' Choice and approve the landlords allowed to engage in this activity.

INITIAL RESPONSE

Already by 1990, it is possible to examine the immediate impact of the Housing Act 1988 on the work of housing associations. Its impact is apparent in the areas of rents, development activity, structures, tenant transfers and other relationships with local authorities.

In responding to the government's new funding regime, the NFHA placed great emphasis on adopting a broad-brush measure of affordability when seeking to influence grant rates. Later this work provided guidance for associations to determine their own policies in the absence of any government or Housing Corporation definition of the concept. The system of 'indicator rents' devised is meant to be a broad indication of affordable rents for different household types, based on a figure of 20 per cent of average net disposable income and taking into account housing benefit eligibility and regional variations in income (Treanor, 1990). A continuous recording system enables the NFHA to analyse new tenants' incomes and household composition. Early experience of rent setting after the Act indicates that, although rents are increasing faster than inflation, they are, in general, being restrained to indicator rent levels rather than rising to the levels implied by the new grant regime. Associations are attempting to hold down rents for new tenants by cutting development costs; reducing on-costs; and, in the case of larger associations, cleverly using tariff arrangements and adopting rent pooling between existing and new tenants to the extent possible under the RSF system (Lupton, 1989). Longer-term problems of inadequate grant levels and a housing benefit system which creates a poverty trap for the working poor remain (NFHA, 1990a), but the real impact of the 1988 Act on rents had not been felt by 1990.

In fact, associations were generally able to deliver the 1989–90 Housing Corporation development programme without altering their client group. They were aided by the favourable tendering climate for rented housing caused by the collapse of the private house-building industry and falling property and land prices in the wake of escalating mortgage interest rates. There were signs of innovative borrowing arrangements for private finance and the adoption of package-deal and design-and-build procurement methods. However, the pattern of development activity was influenced as new building, with its more predictable costs, was favoured over rehabilitation, and special needs schemes were given an uncertain future because of unresolved issues of revenue funding. 'Development drift' (Cope, 1990, p. 296) rather than 'rent drift' was the result of the 1988 changes most evident by 1990.

While associations themselves responded quickly to the new environment, the Housing Corporation did not. Government expenditure plans showed the Corporation's ADP expanding from £815 million in 1989–90 to £1,735 million in 1992–93, and a steady rise in new approvals and thus starts was expected. However, the system introduced by the Housing Act 1988 increased the rate at which houses and land bought are developed into completed homes and, in passing risk to associations, removed the centralised control mechanism of associations having to queue for specific tender approvals. An overspend of £120 million in 1989–90 exposed failures in the Corporation's financial forecasting systems, and was followed in 1990–91 by substantial cuts in allocations from those previously forecast; no cash backing for the bulk of allocations; and the introduction of cash planning targets for each association. The total rented programme of new approvals is expected to be 8,300 in that year compared with the previous target of 19,550. Whether the Corporation will be able to hit its target of 30,000 new rented approvals in 1991–92 is unclear. The Corporation itself believes that the 1990 circumstances are a one-off event (Housing Corporation, 1990b), but it seems that a significant culture change is needed in its organisation. Cash planning targets need to be implemented as an enabling rather than controlling mechanism, with associations themselves taking responsibility for controlling expenditure within agreed objectives (Hoodless, 1990).

The new environment is beginning to influence internal management structures within associations and relationships between them. Within associations there are signs of a gearing up of management structures to respond to the transfer of risk. As the old regime of financial neutrality is replaced by one which emphasises an individual association's asset base, financial and development skills and overall competence, there is a trend towards mergers and 'take-overs' as associations perceive the need to establish the critical mass required to operate effectively. To some extent smaller specialist and new black associations are being assisted by the growth of consortia arrangements providing for co-operation between associations on larger developments, but they face a tough future. One association director has predicted that the aftermath of the 1988 Act will see major variations between the ability of different associations to survive and grow and that 'two dozen or so will expand rapidly and put a distance between them and the rest' (Hoodless, 1989). Losses by home-ownership associations because of the overall collapse of the UK property market may complicate this picture.

The differential opportunities and problems created by the new framework are being accompanied by greater competition between associations, regional jealousies and strains within their representative body, the NFHA. During the late 1980s there was considerable argument amongst associations about the distribution of development funds between the English regions. This became a 'north versus south' debate in the context of reduced programmes and the inclination of government to avoid implementing the conclusions of its Housing Needs Index (HNI) in order to focus investment on lower-cost northern regions and thus increase the number of approvals. For 1990–91 the Corporation and DoE responded to criticisms of its use of the HNI (National Audit Office, 1989) by undertaking a review, but the outcome severely disadvantaged several London boroughs with a high incidence of homelessness and multiple deprivation. The NFHA itself launched a review of its internal representative structure in the light of these regional strains (NFHA, 1990b). Less susceptible to constitutional amendments to hold the 'voluntary housing movement' together, however, are the increasing indications of competitive behaviour among associations and the collapse of zoning arrangements. Some large associations, often but not exclusively from northern England, were accused of predatory behaviour as they moved outside their traditional areas of operation.

One area of activity where there is less predatory activity than might have been expected is in the operation of the Tenants' Choice procedures in the 1988 Act. The legislation was accompanied by a system of voting amongst council tenants in blocks of flats heavily weighted in favour of acquiring landlords by counting abstentions as 'yes' votes (DoE, 1988a). However, from a radical free-market perspective the government tied itself in red tape by the approval criteria for prospective acquiring landlords and the Housing Corporation procedures for the stages of Tenants' Choice. Initially, and before the impact of the Local Government and Housing Act 1989 on rents and standards of service was really felt, council tenants responded by preferring 'the devil they knew', their existing local authority landlord. Some tenants' associations used the threat of the legislation to encourage councils to 'raise their game' and to increase tenants' participation in management without a change of ownership (Fraser, 1988). Associations themselves responded to the legislation by adopting a code of practice which emphasised positive support from tenants (not the rigged ballot system); lack of reasonable opposition from a local authority; and the availability of resources to secure improvements with affordable rents as conditions of their involvement. At a local level many associations entered into agreements, or pacts, with local authorities covering both parties' behaviour under the Act, and often much more (Langstaff, 1989). By the summer of 1990 not one transfer of ownership under this part of the 1988 Act had been completed.

A much more significant activity, and one not fully envisaged before the 1988 Act, was the wholesale, not piecemeal, transfer of ownership of council housing in a particular local authority to a new landlord. This was a course of action adopted by mainly Conservative-controlled rural district councils in the south, in the process establishing a new body, usually a registered housing association, to receive the tranferred stock. Their motivations were mixed – including a wish to escape the implications of the 1989 Act, as well as ideological predispositions towards privatisation. Reactions of tenants were also mixed as they were consulted and eventually voted under the procedures adopted for voluntary transfers (DoE, 1988b). By June 1990 some eleven transfers were completed or underway, involving 60,000 homes; eight had failed at the ballot stage; and twelve had been withdrawn or deferred. Considerable amounts of private finance had been raised and the introduction of a new group of relatively large organisations into the housing association sector had been set in motion.

The immediate post-1988 Act period also witnessed a number of other developments in the relationships between local authorities and housing associations. The government's proposals before the Housing Act 1988 had envisaged ending local authorities' role in funding associations. This was abandoned, but the Housing Corporation was given responsibility for HAG schemework administration on almost all local authority funded schemes. More significantly, the further controls over local authorities' capital expenditure brought in by the Local Government and Housing Act 1989 are expected to severely restrict their support of housing association development, although in the short term some authorities used accumulated capital receipts at the end of the 1989–90 financial year to fund 'off the shelf' purchases by associations before the Act began to bite.

Relationships with local authorities are not confined to HAG funding. Particularly in London, innovative private-sector leasing and development leaseback arrangements were adopted, with associations working in partnership with councils to provide short-term solutions for homeless families in ways which avoid government spending controls. Some local authorities began to plan for the transfer of residential homes to associations. This, together with the work of associations with health authorities in the 'care in the community' field indicated the emergence of a new pattern of 'contract culture' work. It is also extending the role of associations into the area of care and support, previously assumed to be the domain of statutory social services.

More generally, the government's policy of turning associations into the main providers of new subsidised housing focused greater attention on their letting policies and practices. In the mid-1980s housing associations found themselves under scrutiny in terms of the racial composition of their tenants and the effectiveness, or lack of it, of their equal opportunities policies (Niner, 1985). Now, there is also pressure on associations to contribute more to helping local authorities meet the needs of statutorily homeless households and, despite some constraints on their ability to do so (e.g. the dominance of small units in their existing stock), there is accumulating evidence that they are not doing enough (Audit Commission, 1989; Coulter, 1989). Positive action has been initiated to remedy this (Levison and Robertson, 1989; Randall, 1989; Pilkington, 1989).

INDEPENDENCE, ACCOUNTABILITY AND PARTNERSHIPS

Dismantling the role of council housing – or at least severely pruning it to a residual position – and boosting housing associations raises a number of important questions about the role of the statutory and voluntary sectors in national housing policy. These questions have been less often addressed than those related to personal social services (Brenton, 1985), but there are many parallels.

Some of the questions are practical ones. Can associations step up their production to meet all the requirements for affordable rented housing, variously estimated at 100,000 dwellings per year or more? Can they improve the management of substantial numbers of transferred council dwellings? The answer to both these questions is probably 'yes', though not without a build-up period; the injection of substantial public funding; and, in the case of the transfer of the more problematic council estates, the adoption of radical programmes designed to change not just physical but also social conditions. Whether associations are the most appropriate vehicle for the entirety of these two tasks is a different matter, given the expertise of local authority housing departments and the continuing underlying real strengths of their revenue accounts.

But there are also more fundamental questions about the role of the voluntary and statutory sectors which accompany the move to centre stage of housing associations. One question concerns the independence of housing associations from government, and more especially from the Housing Corporation. A lawyer has described associations before 1988 as 'curious entities' in the British welfare state context and referred to the 'peculiarity of the housing association enterprise, as a private trading enterprise in Law, that is wholly dependent on public funds' (Arden, 1983, p. iii). Having been promulgated in the mid-1970s as complementary providers in relation to local authorities – concentrating on housing renewal and meeting special needs – associations have increasingly been regarded by government as the main providers of general needs housing. The growth in their relative position in the 1980s was accompanied by central government bypassing local authorities as their funding vehicle in favour of the Housing Corporation. In the process associations became more influenced by central government priorities as they pragmatically adapted to changes in the emphasis of its policy funding. Furthermore, the growing importance of the Housing Corporation has been accompanied by ever more detailed requirements governing associations' overall behaviour as its monitoring operations have moved past checking legal and fiduciary propriety. Notwithstanding the good practice contained in some of these requirements and their professed intent to encourage self-regulation (Housing Corporation,

1989), it can be argued that these trends have progressively eroded the autonomous, self-governing nature of housing associations – turning them into hired agents of central government, operating as branch offices of the Corporation.

An associated contradiction of government policy is that, while its support for associations has been expressed in terms of favouring diversity, voluntary input and small size, the move to centre stage is eroding the relevance of these attributes. In the 1970s and 1980s the bulk of development activity was carried out by larger associations. By 1990 the 41 associations with more than 3,000 dwellings in management owned more than half the sector's stock. In the early 1990s, as we have seen, the demands of the new mixed funding framework will reinforce the growth of larger associations. As the management of associations becomes more 'businesslike' it will place strains on the concept of voluntary input from their committees and strengthen the power of paid staff.

A second question concerns the accountability of housing associations. Public accountability to the taxpayer is provided by the monitoring function of the Housing Corporation, but other forms of social accountability – to shareholding members, to the local community where associations operate, to elected local authorities and tenants – have become the subject of much debate. Despite Housing Corporation encouragement to associations to consider these issues (Housing Corporation, 1978), and especially to widen their shareholding membership, in 1990 the Corporation was still commenting that 'the image of the self-perpetuating oligarchy remains for many associations. It contains long term risk for their survivial' (Housing Corporation, 1990a). Some associations had developed methods of increasing the representation of local communities in their structures and involving their tenants (Platt *et al.*, 1987), but many had not. Accountability became regarded as the Achilles heel of housing associations as their significance as providers increased. The 1990s began with government seemingly being most interested in accountability to tenants as individual consumers, with both local authorities and associations providing them with information about their performance. At the same time, government also indicated an interest in the collective role of tenants' participation in management (DoE, 1989). These alternative approaches have been described as the 'Marks & Spencer' and 'Co-op' models (NFHA, 1990c).

When the NFHA initiated a debate within the voluntary housing movement about the direction of change associations themselves wished to pursue in the 1990s, issues of independence and accountability were prominent. Independence from government and affirmation of a role which is different from that of the for-profit sector was a recurrent theme in the NFHA's arguments in the 1970s and 1980s. The defence of a distinct position was needed as Labour governments regarded associations as part of the public sector and Conservative ones classified them with private landlords. The NFHA's response was: 'Associations are a creature of neither the public sector nor the private sector. They are governed not by the profit motive, yet they remain autonomous bodies, separate from local and central government' (NFHA, 1982). A distinction between associations and the for-profit rented sector was part of the NFHA's campaign to influence the 1988 Act when the Conservative government sought to couple associations with private landlords as the 'independent rented sector' and the NFHA responded with the argument that they were part of the social housing sector.

By 1990, in seeking to reaffirm associations' independence and to explore mechanisms of achieving public accountability, the NFHA began to make a case for dividing the present functions of the Housing Corporation between two agencies – one to deal with investment and the other to monitor management, particularly services to tenants. It suggested that the registration and monitoring roles of the Corporation could be transferred to a new National Housing Standards Agency, which could also monitor performance in council housing. It also began to argue that a new legal form of incorporation for associations may be desirable. The roles of both the Registrar of Friendly Societies and the Charity Commissioners, who are notionally responsible for the primary registration of associations, have become distant and confusing, particularly when associations seek to widen their activities into home ownership or housing-related neighbourhood activities. The NFHA sought to provide models of greater social accountability for associations to consider and pursue where appropriate.

The NFHA also emphasised both the providing and enabling roles of local authorities, arguing that they should be allowed to provide new homes in numbers matching those envisaged for associations. To make a reality of their enabling role, it stated that local authorities should develop and publish local housing planning strategy statements in co-operation with housing associations, other voluntary bodies and the private sector.

Perhaps the most distinctive feature of the consultation document prepared by the NFHA's working party – established to consider the roles of housing associations – was its emphasis on associations' activities being rooted in the communities they serve. The working party's vision for the end of the decade was that ' . . . the housing association movement will, by the provision of social housing and associated services, and by working in co-operation with others, assist in the creation of healthy neighbourhood communities' (NFHA, 1990c, p. viii). This statement develops a trend established in the 1980s of some associations evolving into more complex multi-purpose agencies, often using funding established for tangential purposes such as employment training to provide non-housing services (NFHA, 1990d). It emphasises their neighbourhood renewal role in partnership with local authorities and other agencies, which developed in the 1980s (Carley, 1990). However, if associations are consciously to align their activities with a wider process of community engagement and empowerment, the statement is only a starting point in exploring the issues involved.

In providing social housing it is the terms of the partnership with central government which will most influence the role of associations in the 1990s. In its fourth term the Conservative government may provide a further stimulus to the privatisation of council housing by a rents-into-mortgages scheme and push both council and housing association rents up to market levels. Its 1988 changes would indicate that it has so far settled for the penetration of market mechanisms into the operation of the non-profit rented sector rather than assisting profit-oriented agencies, either in development or tenant transfer. That trend continues with the Housing Corporation, for instance, indicating that bidding by associations for its 1991–92 programme will be accompanied by cost competition between associations – HAG auctioning, as it has been called – and the avoidance of high-cost areas even if there are high needs there. Given its dominance as a virtual monopoly in providing HAG, the Corporation is in a very strong position to influence development patterns.

Whether this trend will be accompanied by more whole-hearted support for the expansion of the private rented sector is uncertain. At the time of the 1988 Act government spokespersons expressed their disappointment that housing associations were less than enthusiastic about some elements of the package (Trippier, 1989). Of surprise to the government were the results of a study which, in very simplified terms, indicated that greater effectiveness in housing management by associations was accompanied by higher costs than that in council housing (Maclennan *et al.*, 1989). Associations themselves were conscious of the potential for comparable subsidies to be made available to the private sector if they failed to perform (Kilburn, 1989). More recently, one critic from a free-market perspective has argued that associations look more like an extension of council housing than a replacement for it and that their privatisation is necessary if the necessary radical changes in the British housing market are to be achieved (see Chapter 6). If this view prevails then observers in the late 1990s will view present changes as very much a transitional phase.

Housing associations' future under a possible new Labour government after an election in 1992 is also unclear, although some aspects of the party's policy are spelt out (Labour Party, 1989 and 1990). A new form of secure tenancy, with rent setting independent of the landlord, would be introduced. Associations would be subject to a revenue-based subsidy system on a 'level playing field' with council housing. Tenants' Choice and the right to buy would be introduced for all association tenants within a framework of consumerism and for the latter, the duty to replace. Larger associations would be expected to decentralise. What is lacking is an indication of the size of Labour's housing programme and the associations' share of it. Furthermore, how the use of private finance can be combined with rent control and revenue subsidies to the satisfaction of private lenders is unresolved.

CONCLUSIONS

The overall trends of UK housing policy are evident in most other advanced capitalist countries – as governments withdraw from direct provision of rented homes, owner-occupation is favoured by the tax system, markets become unstable, and, as overall output collapses, a substantial minority face a growing housing crisis, with homelessness as the tip of its iceberg (Ball *et al.*, 1988). A narrow focus on the ways in which the role of housing associations has changed could be accused of missing the wood for the trees. Nevertheless, associations have, at least in rhetoric, moved to centre stage by the early 1990s, and this move will have a significant impact on the future of UK housing policy. Whether later in the decade their role will be adapted to that of private-sector organisations; whether they will become key agents of a revitalised social housing sector rooted in the community; or whether they will drop back into the shadows, are remaining questions. The answers to which will have an even greater impact.

REFERENCES

Arden, A. (1983) *Report on Housing Associations to the Greater London Council*, London: GLC.

Audit Commission (1989) *Housing the Homeless: The Local Authority Role*, London: HMSO.

Back, G. and Hamnett, C. (1985) 'State Housing Policy Formulation and the Changing Role of Housing Associations in Britain', *Policy and Practice*, vol. 13, no. 4, p. 397.

Ball, M., Harloe, M. and Martens, M. (1988) *Housing and Social Change in Europe and America*, New York and London: Routledge.

Birchall, J. (1988) *Building Communities: The Co-operative Way*, London: Routledge & Kegan Paul.

Brenton, M. (1985) *The Voluntary Sector in British Social Services*, London: Longman.

Carley, M. (1990) *Housing and Neighbourhood Renewal*, London: Policy Studies Institute.

CES Ltd (1989) *The Social Security Act and Rent Arrears in Housing Associations*, Research Report 3, London: NFHA.

Cope, H. (1990) *Housing Associations: Policy and Practice*, London: Macmillan.

Coulter, J. (1989) 'Homelessness: Has our Performance Been Adequate', *Voluntary Housing*, vol. 22, no. 5, pp. 6–7 and 27.

Department of the Environment (DoE) (1987a) *Housing: The Government's Proposals*, Cmnd 214, London: HMSO.

——(1987b) *Finance for Housing Associations: The Government's Proposals*, London: DoE.

——(1988a) *Tenants' Choice*, London: DoE.

——(1988b) *Large Scale Voluntary Transfers of Local Authority Housing to Private Bodies*, London: DoE.

——(1989) *Tenants in the Lead*, London: DoE.

Foster, C. (1986) 'Private Sector Finance for Housing Associations – Treasury Implications', in HRH The Duke of Edinburgh, *Inquiry into British Housing: Supplement*, London: NFHA.

Fraser, R. (1988) *Tipping the Balance: A Guide to 'Tenants Choice'*, Rochdale: Tenant Participation Advisory Service.

Gibson, M. and Langstaff, M. (1982) *An Introduction to Urban Renewal*, London: Hutchinson.

Hills, J. (1987a) *When is a Grant not a Grant? The Current System of Housing Association Finance*, Welfare State Discussion Paper No. 13, London: STICERD, London School of Economics.

——(1987b) *Finance for Housing Associations: A Response to the Government's Proposals*, Welfare Research Note 8, London: STICERD, London School of Economics.

Hoodless, D. (1990) *Response to 'Into the Nineties'*, unpublished paper given at NFHA Development Conference.

——(1989) 'Regions have had their Day – Hoodless', *HA Weekly*, 3 February.

Housing Corporation (1978) *In the Public Eye*, Circular 3/78, London: Housing Corporation.

——(1985) *Corporate Plan 1985*, London: Housing Corporation.

——(1988) *Tenants' Guarantee: Guidance for Registered Housing Associations on Housing Management Practice*, London: Housing Corporation.

——(1989) *Performance Expectations: Housing Association Committee Members Guide to Self-Monitoring*, London: Housing Corporation.

——(1990a) *Into the '90s: Opportunities and Challenges for the Housing Association Movement*, London: Housing Corporation.

——(1990b) *The Housing Corporation Programme 1990–91*, News Special 39, London: Housing Corporation.

HRH The Duke of Edinburgh (1985) *Inquiry into British Housing: Report and Evidence* (two volumes), London: NFHA.

Kearns, A. (1988) *Affordable Rents and Flexible HAG: New Finance for Housing Associations*, Discussion Paper 17, Glasgow: University of Glasgow Centre for Housing Research.

Kilburn, A. (1989) 'Stop Crying: Start Trying', *Roof*, 14, no. 2, pp. 26–7.

Kirby, A. (1981) 'The Housing Corporation, 1974–79: An Example of State Housing Policy in Britain', *Environment and Planning*, vol. 13, pp. 1295–1303.

Kleinman, M. (1987) *Private Finance for Rented Housing: An Analytic Framework*, Discussion Paper No. 18, Cambridge: Department of Land Economy, University of Cambridge.

Labour Party (1989) *Meet the Challenge: Make the Change*, London: Labour Party.

——(1990) *Opening Doors: Labour's New Strategy for Housing*, London: Labour Party.

Langstaff, M. (1984) 'The Changing Role of Housing Associations', *The Planner*, vol. 70, no. 5, pp. 25–6.

——(1988a) 'Chosen to be Chosen: Housing Associations and Tenant Transfers', *Housing Review*, vol. 37, no. 5, pp. 173–5.

——(1988b) 'Finance for Repair: Building Maintenance into the Plan', *Voluntary Housing*, vol. 21, no. 10, pp. 20–4.

——(1989) 'A Pact in Time. . .Saves: Housing Association – Local Authority Agreements', *Housing*, vol. 25, no. 6, p. 6.

Levison, D. and Robertson, I. (1989) *Partners in Meeting Housing Need*, London: NFHA.

Lupton, M. (1989) 'Grasping the Nettle', *Housing*, vol. 25, no. 9, pp. 10–13.

Maclennan, D., Clapham, D., Goodlad, M., Kemp, P., Malcolm, J., Satsangi, M., Stanforth, J. and Whitefield, L. (1989) *The Nature and Effectiveness of Housing Management in England*, London: HMSO.

Morton, J. (1989) *The First Twenty-Five Years*, London: Housing Corporation.

National Audit Office (1989) *Department of the Environment: Housing Association Grant*, London: HMSO.

National Federation of Housing Associations (NFHA) (1982) *Housing Associations: Their Contribution and Potential*, London: NFHA.

——(1987a) *Standards for Housing Management*, London: NFHA.

——(1987b) *Rents, Risks and Rights: NFHA Response to the Government's Proposals*, London: NFHA.

——(1990a) *Paying for Rented Housing*, Research Report 12, London: NFHA.

——(1990b) *Regional Structures: A Consultation Paper*, London: NFHA.

——(1990c) *Towards 2000: The Housing Association Agenda*, London: NFHA.

——(1990d) *Getting Involved: HAs and Community Involvement*, London: NFHA.

Niner, P. (1985) *Housing Association Allocations: Achieving Racial Equality*, London: Runnymede Trust.

Paris, C. (1989) 'Low-Cost Social Rental Housing: Are there British Lessons for Australia?', paper given in the Sir Robert Menzies Centre for Australian Studies seminar series.

Pilkington, E. (1989) 'The Toughest Challenge', *Roof*, vol. 14, no. 6, pp. 25–7.

Platt, S., Piepe, R. and Smythe, J. (1987) *Heard or Ignored: Tenant Involvement in Housing Association*, London: NFHA.

Randall, G. (1989) *Tackling Homelessness*, London: NFHA.

Shelter (1987) *Homes for the Rich and not the Poor? The Government's Plan for Housing Associations*, London: Shelter.

Treanor, D. (1990) *Housing Association Rents*, London: NFHA.

Trippier, D. (1989) 'The Housing Act 1988: Looking to 1990', speech given at Institute of Housing Seminar.

3 Building societies
Builders or financiers?

Douglas Smallwood

In the 1990s building societies will have a vital role to play in financing the implementation of housing policy. At the same time they are unlikely to become builders on a substantial scale. This chapter sets out the reasons for these assertions. It is based largely on the experience of the Halifax Building Society, although it also considers some of the work of other societies. The track record of building societies as financiers and providers of social housing during the 1980s is examined. The possible development of this role into the 1990s is assessed in the context of the radical changes currently taking place in the work of building societies, particularly the directions into which they are choosing to diversify their activities.

BUILDING SOCIETIES AND FINANCE FOR SPECIAL HOUSING MARKETS IN THE 1980s

Many housing professionals and commentators debate long and hard about the meaning of the term 'social housing'. Building societies do not concern themselves with this debate. They perceive that housing policy and legislation provides them with business opportunities. As these relate to new markets which require specific product research and development, the societies often refer to them as 'special housing markets'. This difference in terminology between 'social housing' and 'special housing markets' is important because it illustrates that building societies have become active in areas of housing other than traditional owner-occupation for business rather than altruistic reasons. This is not to argue that their mutual status is irrelevant. Indeed this mutuality means that they have a vested interest in contributing to a healthy market across all types of provision and tenures, as such a policy is clearly of major interest to their members. For example, by funding major urban renewal schemes the societies are assisting initiatives of fundamental social importance.

The business case for diversification into special housing markets was based on the realisation that although societies had very successfully promoted owner-occupation, much more needed to be done in the margins of owner-occupation and in the rented sector. By 1980 owner-occupation had grown to well over 60 per cent, considerably higher than most other European countries, and many felt that the mortgage market would peak in the near future. In addition it was foreseen in the early 1980s that competition from other lenders would increase, and indeed the societies' share of the market slipped from 92 per cent in 1981 to 72 per cent by 1989. In the light of these market trends, societies lobbied for new powers to enable them to diversify into new markets. A series of Building Societies Association reports argued for a 'level playing field' with the clearing banks in the personal finance sector and for societies to be able to offer unsecured loans and a full home-buying service including estate agency, insurance and conveyancing. A number of these aspirations for the future of the industry were reflected in the new legislative framework in the form of the Building Societies Act 1986 (see Boleat, 1990).

However, societies did not need changes in legislation to enable them to provide some finance for new housing markets. A few societies recognised a potentially significant new market for finance for housing provision which lay beyond traditional owner-occupation. To a large extent the opportunities were provided by the 1980 Housing Act, which resulted in many housing associations working with building societies for the first time. First, registered non-charitable housing associations were enabled to obtain Housing Association Grant for a range of low-cost home-ownership initiatives – improvement for sale, shared ownership and leasehold schemes for the elderly. Second, the Act introduced assured tenancies as a new form of tenancy. These tenancies enabled associations to exercise some control over their income, because rent levels and reviews were not determined independently by the rent officer. As a result both the associations and private financiers were able to have more certainty that rental income during the life of the mortgage would be sufficient to support the loan.

Those societies which chose to lend in these special housing markets undertook their market planning in the same way as they approached any new area. They considered and segmented the market and established that the lending potential was large enough to justify committing the resources to develop and deliver the financial products which were required.

Finance for housing for the elderly demonstrates the segmentation approach. By the mid-1980s some societies had developed a range of approaches to financing housing for the elderly:

1 the provision of development finance to the providers of sheltered housing both for rent and sale;
2 working with housing associations in the provision of rental schemes for the frail elderly by developing special funding mechanisms in order to reduce the subsidy requirement of projects;
3 supporting 'care and repair' and 'staying put' projects again by the development of special products, mainly maturity loans; and
4 developing mechanisms to address a market segment. (A particularly good example is the Sundowner scheme developed by Coventry Churches Housing Association. This scheme addresses the needs of elderly people who require purpose-built accommodation but cannot afford traditional sheltered housing for sale. The financial package which was developed by the Association in conjunction with the Halifax is crucial to this initiative.)

Private finance for rented housing is a second major example of segmentation. Societies identified that the market potential lay in working with registered housing associations who were establishing new organisations in the form of unregistered associations to develop rented housing on an assured tenancy basis. This market was addressed by developing low-start products which enabled mortgage repayments to match the projected rental income streams. When the Halifax placed the first index-linked stock issue for this market in October 1985, they succeeded in providing a particularly low-start product designed specifically for rented housing.

The importance of this approach to social housing cannot be over-emphasised. It 'broke the mould' of equating low-cost rented housing provision with 100 per cent public expenditure. The initiatives addressed all tenures and needs groups. They also demonstrated the key importance of a number of principles for housing policy and its implementation:

1 Partnership – projects demonstrated the principle of partnership not only between the society and the borrower, but particularly with local authorities. They helped assemble sites, resolved planning issues and were instrumental in obtaining a variety of types of capital subsidy which were essential for project feasibility.
2 Redefining the rented sector – housing was provided at fair rent, affordable and market rent levels. This resulted from the fact that our borrowers were not exclusively registered housing associations seeking to provide housing at fair rents. They demonstrated that rented housing was not necessarily a second-class form of tenure but one which potentially could attract a range of customers.

3 The borrower – schemes proved that small as well as large organisations could successfully use building society finance to provide housing.

4 Mixed funding – schemes from the early 1980s used low-start private finance alongside public subsidy. As a result, they pre-dated by several years the Housing Corporation's mixed-funding programme where low-start private finance is used in conjunction with Housing Association Grant. Low-start private finance reduces the amount of public subsidy required, but most importantly these schemes also demonstrate that adequate levels of subsidy support are essential.

5 Mixed tenure – a number of schemes provided for a range of tenures with profits from properties developed for sale providing an additional capital subsidy for the rental element.

6 Pump priming – projects demonstrated that limited public subsidy could draw in the private sector not only as partners in the initial scheme, but as the major players in the development of other projects in the same locality.

BUILDING SOCIETIES AS PROVIDERS OF SOCIAL HOUSING IN THE 1980s

Societies have achieved little as housing developers in comparison with their work in financing housing development. However, there has been a range of approaches to housing provision. A number of the larger societies have set up housing subsidiaries to purchase and develop land. Woolwich Homes has tackled both ends of the housing market – starter housing and sheltered accommodation. Their perception is that there is a gap in the market which can be filled particularly where housing development is linked to specialist funding products. The Woolwich has been at the forefront of developing equity mortgages whereby the society retains an equity stake in the property, thus providing access to owner-occupation at a significantly reduced cost to the occupier. For a number of reasons no society has yet been able to make the product available in significant volume, but the Woolwich has taken the opportunity of targeting their product to some of the schemes developed by Woolwich Homes.

The stated aims of the Nationwide Housing Trust and Anglia Housing Association are to fill a gap in the market between the public sector and the conventional private sector. In 1988 the Nationwide Housing Trust celebrated the sale of its 2,000th home, and the publicity material emphasised the importance that Nationwide Anglia attaches to the Trust's role.

The Northern Rock Housing Trust had 21 projects in the course of development by the end of 1988. Interestingly, in an article in September

1987, the General Manager and Secretary of the Society referred to the fact that these are early days in what was described as the transformation of building societies 'from mere financiers to developers' in this market (*Building Societies Gazette*, 1987).

As well as direct development, there are also a few examples of societies acquiring equity stakes in companies involved in housing development and management. For example, the Halifax and Nationwide Anglia building societies entered into a partnership with the Lovell Group in the forming of Partnership Renewal of the Built Environment (PROBE) Limited, which aims to provide the development expertise, management capability and funding packages necessary for the reclamation of large areas. Its role is that of catalyst in urban regeneration. It is a vehicle through which the societies provided funds targeted to tackling the problems of urban decay.

The Quality Street initiative, launched by Nationwide Anglia in 1987, is a further example of the range of approaches societies have already made to the process of the provision and management of social housing. At the launch, the Society announced its plan to invest £600 million in the company over five years in order to provide 40,000 homes. Its aim was to provide good-quality rented housing for a range of needs and demands in order to create a fluidity within the market. The stated objectives of the initiative included:

1 The promotion of choice of tenure;
2 A belief that the growth in owner-occupation may have reached optimum levels in various sectors. The Society has quoted the example that the UK has the highest proportion of owner-occupation in younger age-groups (up to age 34) than any other advanced industrial nation; and
3 A response to the demand for private rented housing from job movers, divorcees, those finding the high costs of owner-occupation a burden, and elderly people no longer wanting the responsibility of repairs.

The relatively small number of these initiatives reflects basic commercial logic. Why should building societies move into housing development and management, areas far removed from their traditional areas of activity, when they can choose to diversify into new markets which are directly related to the mainstream mortgage business? The existing strengths of the organisation can be used to take advantage of such opportunities as the provision of insurance and personal banking services. By contrast, how can building societies expect to compete successfully with major housing developers in their existing markets?

In addition, as the housing market of 1989–91 has amply demonstrated, housing development is a high-risk business, and societies perceive that significant activity as developers might undermine the perception of financial stability which is a hallmark of their industry. Societies are constantly aware of the need for investors to have confidence in the financial stability of the organisation. The needs of the investor are paramount – the industry's membership comprises six times as many investors as borrowers.

NEW OPPORTUNITIES FOR HOUSING FINANCE IN THE 1990s

It would be an exaggeration to claim that all housing policies initiated in recent years have provided potential new areas of business for building societies. Nevertheless, a central feature of housing policy has been to promote means by which private funding could be used for social housing. In addition, not all initiatives will achieve the levels of investment by building societies as might have been envisaged. For example, the government's attempt via the 1988 budget to provide a 'kick start' to new investment in private rented housing through the Business Expansion Scheme has had some success, but it has not so far provided a vehicle for the investment of building society funds. In the first two years of the scheme, 8,200 homes have been provided by 188 companies. In that time, £461 million has been raised by share issues. In addition, the scheme has resulted in the provision of social housing, as well as catering for the market rented sector (see Joseph Rowntree Foundation, 1991). Investor interest has been patchy, with companies that are seeking to provide social housing struggling to offer potential investors a projected return significantly greater than could be expected from a building society deposit account. A striking feature is that only a handful of these companies have sought to gear up the shareholders' funds by loans from building societies in order to provide additional rented homes.

The 'rent-into-mortgage' pilot projects are yet to be fully established, but it is doubtful whether a scheme which is by definition operating at the margins of owner-occupation will prove attractive to private financiers who are currently facing record levels of mortgage arrears and repossessions. Nevertheless, it should be noted that those societies which were providing finance for special housing markets in the 1980s were already often working on the initiatives propounded as key elements of new housing policy in the Housing Act 1988.

Some housing policy initiatives

1 Private landlords and assured tenancies

Assured tenancies were introduced by the 1980 Housing Act and extended in the Housing and Planning Act 1986. They provide an opportunity for private landlords to let properties at market rents; the legislation, however, has had very limited success. By April 1987 there were 217 organisations approved to offer assured tenancies, but only 53 of them were letting properties and the total number of lettings was only 2,998. Housing associations registered with the Housing Corporation were ineligible for approved landlord status for the purposes of offering assured tenancies. The large majority of organisations letting dwellings using assured tenancies were unregistered associations set up by existing housing associations specifically to offer assured tenancies. These organisations demonstrated what could be achieved by private landlords using assured tenancies. Both unregistered associations and private landlords operate without the benefit of Housing Association Grant; they provide assured tenancy lettings; and they are eligible for a similar range of grant support, including City Grant and financial assistance from local authorities under Section 24 of the Local Government Act 1988. The key difference, of course, is that they also operate within very different tax regimes. A private landlord is simply unable to obtain an adequate return on investment in social rented housing when taxed on rental income and capital gains.

2 Housing associations and mixed funding

By the mid-1980s societies had supported many projects which used a mix of public subsidy and low-start private finance. By 1990 this has been extended to become a key feature of the Housing Corporation's Approved Development Programme. In addition, in November 1990 the Housing Corporation announced a projected £1.75 billion private-finance requirement for their programme over the next three years, together with a substantial additional demand for wholly privately financed activity, including the transfer of stock from local authorities.

3 Tenants' Choice

Again, societies have supported major initiatives since the mid-1980s. Examples from the Halifax portfolio include the purchase and improvement of local authority and British Coal-owned properties by tenants' co-operatives with management support from established housing associations. This developed from 1989 into the funding of voluntary transfers, whereby the ownership of the whole of the local authorities' housing stock is transferred to a housing association formed specifically for that purpose. By early 1991, 13 transfers had been completed, requiring low-start funding of more than £1 billion from building societies and banks. The Halifax and National Westminster Bank have been at the forefront of this market, and they anticipate continuing demand for funding for transfers both wholesale and on a trickle basis.

Of course, the analogy between this work and Tenants' Choice under the Housing Act 1988 is incomplete. In these examples there has been a consensus between the local authority and the new landlord. I anticipate that it will be more difficult to achieve viable schemes of this type where the local authority does not actively support the proposals. However, the above examples demonstrate the willingness of a building society to fund the purchase of tenanted housing by a range of types of landlords wherever it can be satisfied that adequate security is provided for the advance and that the borrower has the financial resources to repay the loan and can demonstrate that the properties will be managed and maintained to a high standard.

4 Housing Action Trusts

I believe that many of the principles underlying this housing finance programme in the 1980s are echoed in the Housing Action Trust strategy. The partnership approach between a number of organisations, the use of public subsidy as 'pump priming' drawing in private finance, and the importance of a comprehensive planning approach to urban regeneration have all been illustrated. The issue of whether a Housing Action Trust is the most appropriate vehicle to tackle areas of acute deprivation should not deflect from the importance of these principles.

5 Home improvements

The provision of finance by a building society for home improvements both to individual borrowers and for initiatives such as enveloping and block improvements is well documented. Towards the end of the 1980s a major complication arose from the Consumer Credit Act 1987 whereby loans of less than £15,000 for improvement were not readily available from most societies unless the property was already in mortgage to the society and the funds were provided by way of further advance. However, an Exemption Order in May 1988 enabled 'lending to proceed wherever an "agency service" is in operation which provides free impartial advice concerning the improvement proposals to the borrower'.

So, as some building societies are already working in the markets promoted by the legislation, the opportunities for them therefore relate to market penetration and product development rather than to diversification. The 1988–89 year marked a substantial increase in demand for building society finance for special housing projects; the size of the Halifax's portfolio, for example, more than doubled. By the end of 1989 the Housing Department was administering more than 600 projects with mortgage advances approaching £1 billion. The level of activity plateaued in 1990 as a result of reduced demand from housing associations caused by the Housing Corporation's 'cash crisis' and by the reduction in low-cost home-ownership provision, itself a result of the general problems surrounding house prices and disposals. Halifax is committed to providing finance for all housing tenures and certainly is achieving as much as any other financier in the rented housing market. It has been at the forefront of innovation both by developing specialist low-start products and working in partnership with enablers and providers. Yet even this portfolio is relatively small when compared with the likely scale of the demand for private finance for rented housing alone over the three to five years.

Specialist low-start private finance

The question of the capacity of private finance is not simply one of the quantity of the funding required. If it were, we could point to the number of potential financiers that exist and the size of their mortgage portfolios. The type of funding required is as important as volume because rented housing and low-cost home-ownership projects invariably require specialist products with a low-start repayment mechanism which can be serviced from rental income. The private sector certainly has no difficulty in providing the quantity of finance required, but it is unrealistic to presume that it is able to provide low-start finance over a 25–35-year term in sufficient volume to support the new programmes. Building societies obtain funds both 'over the counter' (i.e. retail sources) and from the money markets (i.e. wholesale sources). The difficulties in using retail funds can best be illustrated by the simple point that by making funds available on a low-start basis, the financier is receiving less interest from the borrower during the early years of the mortgage term than is being paid to the depositer. As a result, products such as deferred-payment mortgages can only be made available so long as they remain a very small part of the funder's mortgage portfolio. When using wholesale funding, societies will limit their risks by 'matching' the type of mortgage required by the borrower (e.g., low-start finance) to a similarly structured instrument from the wholesale markets (e.g., index-linked stock). Funds are available on a 'matched' basis so long as institutions wish to invest in the stock. Invariably there are only a very limited number who wish to do so, particularly when a long-term arrangement is sought. Clearly if the Halifax Building Society has difficulty in obtaining a constant supply of long-term, low-start finance from institutional investors, this does not bode well for any other private financier.

Therefore it must not be presumed that private finance will be available at the time, in the quantity and of the specialist nature which our borrowers will ideally be seeking. There has to be an acceptance of this fact by government, enablers and our borrowers. The Housing Corporation's mixed-funding programme should take account of it. Grant levels need to be set which enable housing associations to utilise conventional private finance as well as 'low-start' products. Similarly, purchase prices for stock transfers need to be agreed on the basis that conventional mortgage finance rather than long-term, fixed-interest or low-start finance will be the source of funds.

Most important, it needs to be recognised that private finance for rented housing is not synonymous with low-start finance for rented housing.

The Halifax has provided conventional funding for several projects of affordable rented housing. What is required is a professional but innovative approach by all partners to the project, adequate levels of subsidy and local authority support.

Implications of a competitive financial market

To date, the City has shown comparatively limited interest in these markets. However, clearing and foreign banks and indeed a larger number of building societies are becoming involved. These new opportunities have come at a timely moment. Local authority borrowing requirements which have for so long been a major source of business for the societies are severely reduced and central government has repaid £28 billion of debt during 1988–91 as part of a longer term fiscal policy. However, some of these traditional and attractive markets are now returning. For example, the government is currently projecting a significant public-sector borrowing requirement for 1991 and 1992. As a result, the interest of the City in financing social housing may well prove to be transitory, or the market may become one in which they play irregularly. In the meantime, borrowers will be able to consider a variety of potential financiers.

The consequence is of course that each player will be striving, to use the marketing jargon, for their 'distinctive competence'. There may well be two aspects to this. The first is product innovation and the need for funders to offer a comprehensive range to enable them to meet the requirements of their customers. The second is customer service, particularly as these markets are so specialised. The standard of customer service both pre- and post-sales is just as important for this market as for any other. This is a particular challenge in a period of rapidly increasing demand for new financiers who are in the process of developing administration systems for this specialist area of work.

OPPORTUNITIES FOR SOCIETIES AS PROVIDERS OF SOCIAL HOUSING IN THE 1990s

Although there are good business reasons for societies to choose to limit their activities as housing developers, there remains clearly a huge demand for the provision and management of social housing. The housing association movement has been producing 15,000-20,000 homes per year for the past ten years. The Housing Corporation's most recent projections are for a programme of 30,000 homes in 1991–92. This still is only on a par with the output of the movement in the early 1970s. With Treasury projections indicating that public expenditure on housing provision will remain fairly constant for the foreseeable future, and with housing waiting lists in the average large town continuing to grow and typically comprising more than 2,000 applicants, there remains a massive need and potential for other providers of social housing.

There must also be some doubt about the ability of all but a small number of housing associations to achieve the level of provision that government envisages, even if substantially more subsidy was available to support their programmes. Housing associations generally have limited capacity and desire for major expansion. Their capacity is limited by the size of their current activity. At present they account for only 3 per cent of the housing stock. Local authorities manage ten times the number of dwellings owned by housing associations. Associations traditionally have seen their role as supplementary and complementary to the local housing authority. The majority do not want to become the major providers of social housing or take over local authority stock. They have other aims, usually related to small-scale, localised, community-based work. It is quite possible that the movement will only be able to achieve a much expanded level of provision through the emergence of about twenty very large associations developing on a national or regional basis. Would such a scenario be attractive or even acceptable to government? Is it not the voluntary, responsive and locally based image of associations that makes them acceptable across the political divide?

Despite this potential for other landlords, the fact remains that private landlords are unable to make a significant financial return to justify

investment in social housing. It is most unusual for landlords other than
local authorities or housing associations to obtain capital subsidy support
for housing provision despite the fact that the 1988 Local Government Act
enables local authorities to provide assistance to private landlords. Without
subsidy, the cost of occupation for a tenant would be higher than the
outgoings of an owner-occupier of an equivalent dwelling with a mortgage.
At the launch of the Quality Street initiative mentioned above, Nationwide
Anglia emphasised the need for both capital subsidy and an overhaul of the
housing benefit system, if the company were to be able to address the
low-cost market. Practical experience has served to illustrate their point.
Quality Street, four years after its launch, has provided only a fraction of
the 40,000 homes it projected at its launch, and much of that has not been
social housing. So although the market for high-cost rented housing for
those who choose to rent rather than purchase (for whatever reasons) might
be serviced, building societies would not be able to compete as providers of
social housing until there is adequate subsidy support either through the
form of capital grants or by the fundamental review of housing benefit
levels and the taper system (Coleman, 1988).

At the same time, building societies probably will have the opportunity
to become project co-ordinators on a more significant scale in the 1990s.
There are three sets of circumstances where this is most likely to occur:

1 Probably most important, developers recognise that the key to the
   provision of low-cost housing lies in the availability of specialist
   financial products and the potential of building societies to have access
   to land owned by local authorities which might not usually be offered to
   a developer. They recognise the need to work with building societies
   where these special markets are being addressed. As local authorities
   take on the mantle of enablers and direct through the planning and other
   processes the implementation of their local housing strategies, builders
   will need to work closely with societies from the inception of projects in
   order to meet the briefs of local authorities.
2 Where local authorities or other public bodies dispose of major sites and
   wish to secure the provision of a significant number of social housing
   units, it is quite likely that several developers and housing associations
   may be involved. For example, the consortium is likely to include
   housing associations which specialise in different sectors (e.g. specialist
   housing for the elderly versus low-cost home ownership). In these
   circumstances, there is great attraction to the local authority in the
   possibility of being able to deal directly with only one organisation
   rather than with a large number of developers. The attraction of working
   with a building society in these circumstances is partly the securing of

the necessary funding and partly the benefits to be derived from using the corporate image of the society in the promotion of the project.

3 Difficulties in the relationship between local authorities and housing associations will undoubtedly become more widespread in the 1990s. To date, housing associations have seen their role as complementary and their provision as supplementary to that of the local authority. Now, not only do housing associations have greater available subsidy support than local authorities for the provision of social housing, but potential conflict exists in areas such as Tenants' Choice where the housing association may often be torn between its own strategies, the demands of prospective tenants, and the policy of the local authority.

Most recently, the Prime Minister, at the Conservative Party Local Government Conference in March 1991, raised the possibility of competition in the delivery of housing management services being introduced for local authority-owned housing. Under such a scenario, the most likely alternative managers are housing associations, so that such a policy could accelerate a move towards local authority housing departments and housing associations being seen as competitors rather than complementers.

BUILDING SOCIETIES AND DIVERSIFICATION

Of course, the response of building societies to new legislation and housing policy which requires private finance will only partly be conditioned by their experience in housing development and finance to date. Housing has to be seen in the context of the key debates currently facing the industry (see Kleinwort Grieveson Securities, 1988).

The primary concern of societies will remain the protection of their share of the mortgage market for home-owners. The importance of this market was emphasised by the Building Societies Act 1986. This Act prescribed that 90 per cent of lending must be in this market, and although this percentage has subsequently been amended to 75 per cent, there is no doubt that the prime business of building societies will remain the provision of mortgages for home-owners. The sheer size of the market also demonstrates its key importance. There has been a huge growth in demand in the 1980s – an average of nearly 20 per cent per year – with the result that societies have grown on average by 17 per cent per annum over this period. Societies advanced more than £35 billion for home purchase in 1987, a figure which had grown from less than £10 billion in 1980. The Halifax alone advanced more than £10 billion to house purchasers during our financial year 1988–89 and still advanced more than £8 billion in 1990–91 during one of the most difficult housing markets on record.

The spectacular growth during 1988 is unlikely to be repeated in the foreseeable future. Market share has become fundamentally important in a contracting or static market and it will become increasingly difficult to achieve. As a result, lending margins will come under more pressure and societies will concentrate on trying to secure sources of mortgage business. This has led a number of societies to adopt the strategy of acquiring estate agents. The scope for selling mortgages through agents is huge, and, in principle, high street insurance brokers and even retail outlets are also potential points of sale of mortgages. These market conditions are also leading societies to reassess their sales strategies. Branch networks have been reviewed, programmes of computerisation and installation of ATMs (e.g. card-cash machines) brought forward, and the importance of customer service emphasised.

Problems in the core business area of mainstream mortgage lending are not confined to lending volumes or market share. The 1990–91 year has witnessed, for really the first time in the history of the movement, significant levels of provisions for losses in building society accounts. A particularly striking example for social housing is the £35 million provision made in 1991 by the National and Provincial Building Society in respect of its lending for self-build housing. Alliance and Leicester has provided for losses of £33 million in its commercial lending, £14 million for mainstream mortgages and £8 million for personal loans. This experience is not uncommon among other societies. As a result, a key issue running through the heart of the movement is the need to secure good-quality lending and to monitor and control lending already on the book, particularly commercial lending. In this environment, social housing must be a secondary priority for building societies, particularly as it is still regarded as potentially a risky area of business by many societies and by the Building Societies Commission. Nevertheless, many societies continue to recognise the need to diversify. This is partly the result of a basic marketing principle that companies should seek to diversify from positions of financial strength in mature markets. However, it is also particularly the result of the Building Societies Act 1986. Prior to the Act, societies had to lend against the security of freehold or leasehold estate. The Act enables them to respond to the demands of customers for a wider range of services, including unsecured lending, insurance services, banking services, subsidiary operations including estate agency, and the purchase and development of land. Societies are having to choose between areas for diversification. They simply do not have the management capacity in the short term, nor is it necessary for business reasons, to diversify into all potential markets. The importance of this fact for housing is that the logical diversification for many societies is in the provision of more services to the existing customer base rather than in new housing markets.

Unit trusts and insurance are good examples. Societies have always regarded the unit trust industry with suspicion, viewing it as a competitor for personal savings. However, it is considered likely that there will be an increased interest in equity-based investments as unit trusts can now be held in personal equity plans. Societies could be able to take advantage of the personal-sector interest in unit trusts by the acquisition of unit trust groups. Insurance products are major examples of how societies can harness the cross-selling opportunities arising from mortgage business. Life insurance, home insurance and indeed all personal insurance requirements are market opportunities of which societies can take advantage. Many societies are operating in these markets in a substantial way, often in a joint venture with an existing insurer. Of course, societies might decide to buy into the insurance business as this is the quickest route into insurance without the operational and management costs of building up an operation from scratch or dealing with a joint venture partner. A good example of this approach is the recent move by Britannia Building Society effectively to acquire FS Assurance.

The Alliance and Leicester Building Society's purchase of Girobank is another mechanism whereby a society is choosing to diversify by seeking to expand both its customer base and the range of personal financial services it can offer. The society has access to 20,000 post-office counters and 2.5 million customers and has acquired a current account banking service (Slaughter, 1989).

A further example of diversification, and again one with major implications for the future of societies as providers of finance for special housing markets, is the question of conversion to public limited company (PLC) status. At the time of writing, only Abbey National Building Society has converted to PLC status, but there are suggestions that other societies, including National and Provincial, may follow this lead. The background to the debate is the 1986 Building Societies Act which gave societies the right to offer a wider range of services but failed to produce the 'level playing field' with banks which the societies were seeking. For that reason, Jim Birrell, Chief Executive of the Halifax, called for major revisions to the legislation at a speech to the Midlands and West Association of Building Societies in October 1990. He argued that the regulation of the activities of any financial institution should be based upon capital requirements, and pointed out that banks are able to operate in any market provided that they can meet the capital requirements of their regulators. By contrast, building societies can operate only in the framework described by the Building Societies Act (*Mortgage Finance Gazette*, 1990). A simple example is quoted in a recent report by Global research group at UBS Phillips and Drew. This suggested that the biggest threat for building societies is the legally imposed restrictions of their business activities and, in particular, the 40 per cent maximum limit on funding wholesale markets. It suggested that this could put them at a distinct competitive disadvantage with the banks in the medium term (UBS Phillips and Drew, 1991). The second example of the sort of difficulty caused by the prescriptive nature of the legislation is in the syndication of loans. A bank is able to share the security for a loan with other lenders. By contrast, a building society has to have a first fixed charge over the security as a prerequisite to lending. Therefore, for larger commercial transactions, including loans of, say, more than £10 million for social housing, banks can effectively reduce their risk by taking only a portion of the proposal onto their books. As pricing is essentially a function of the assessment of risk, they are often able not only to participate in loans which are too large for an individual building society to fund, but also to price the transaction more competitively than would be possible for a building society.

Some commentators believe that the capital which will be raised by conversion will be crucial for financing future development and diversification. However, some societies take the view that there are substantial benefits to be derived from the status of a mutual organisation. They believe that the Building Societies Commission may relax or remove the most onerous of the restrictions on the activities of societies. They also believe that there are adequate alternative means of raising the required capital, particularly by innovation in fund-raising from wholesale sources.

I believe that this issue is very important for the future of building societies as major providers of finance for social housing. At the end of the day, a plc's business plan needs to be geared first and foremost towards profitability in order to meet the expectations of shareholders. Mutual organisations clearly must be financially viable, but the social importance of the work they undertake is relevant to their members; otherwise the members would place their funds in a bank. It is reasonable to suppose that the provision of finance for social housing will not be as profitable for lenders as other areas of their work. As with mainstream mortgage business, I am quite sure that activity in this market by both clearing and foreign banks will be dependent on commercial market forces. The inevitable conclusion is that social housing requires substantial providers of private finance other than banks if it is to secure the consistency and quantity of funding which is so important for the implementation of housing policy and programmes.

CONCLUSION

Just as the performance of the mainstream mortgage market is largely dependent on social and economic policies beyond the control of building societies, so the future of private finance for social housing is in no small part dependent upon the action of housing policy-makers and implementors. Will they be able and willing to grasp the opportunities which the legislation provides? The concern relates to the ability of organisations to adapt to change. As noted above, building societies actually sought change in their markets and in the legislative framework in which they work. They saw that change was essential in order to enable them to retain a key position as financiers. They recognised that change was inevitable, but they also recognised the opportunities that existed. In the housing field, many questions remain:

1 Most important, will all the parties to the process of enabling, developing and managing housing succeed in working together in partnership? This requires a subtlety of approach which has to date only been achieved as an exception rather than as a rule. This is no doubt because true partnership requires the partners to compromise as well as obtain benefit from the process.
2 Will housing associations become major providers of social housing or will they ultimately choose to continue to be complementary and supplementary in their approach?
3 The demand for private finance for stock transfers will, of course, be largely determined by whether tenants choose to opt for a new landlord. The evidence to date is inconclusive. Fewer than 50 per cent of voluntary transfer proposals where a ballot has been held to date have achieved a 'yes' vote.
4 Will sufficient public subsidy be made available? Experience demonstrates that private finance for social housing must be used in conjunction with subsidy. Treasury figures suggest that public expenditure on housing is not set to increase, but is being redistributed between local authorities and housing associations.
5 Will local authorities have the desire and resources necessary to undertake a new role as enablers across a wider range of housing needs than has been their traditional concern?

The political climate is one factor which is not included in this list. This is because at the time of writing there does seem to be a political consensus about the key issues which have implications for the role of private finance for social housing. In particular, the Labour Party's housing spokesperson, Clive Soley, welcomes the expansion in the work of housing associations and argues not against the introduction of private finance into their mainstream development programmes, but for ensuring that grant levels are adequate to enable rents to remain affordable. He envisages an expansion to the Tenants' Choice scheme by providing similar opportunities to tenant in the private sector. The two areas of housing policy where currently there is most debate – the treatment of mortgage interest tax relief for owner-occupiers and the future of local authorities as providers of housing – do not relate at least directly to the future of the availability or use of private finance for social housing.

Of course in some ways I welcome this consensus. Yet at the same time I believe that the debate on housing policies should be centred on questions relating to private finance. The fact that they are not illustrates that much of the debate fails to address the crucial policy issues. A debate which centres on Tenants' Choice, the right to buy, and possibly rent-into-mortgage makes little contribution to the key objectives of achieving an adequate level of social housing provision and improving our existing housing stock. Such a debate is concerned purely with the transfer of ownership. It may produce choice for tenants, but what about choice for those people on housing waiting lists? At the same time, a strategy which emphasises the role of local authorities as providers or the public purse as the means of delivering social housing is unlikely to result in the provision of sufficient numbers of homes. This is because housing issues are clearly not at the top of any political agenda. It is difficult to conceive that they will be, as long as more than 90 per cent of the population is well housed. This political reality must be reflected in the amount of public-sector resource that can be committed to social housing. Surely the evidence of the past decade is proof enough. There has simply been a continuous decline in the supply of social housing.

As a result I believe that building societies have a crucial role to play in tackling our major housing issues. The changes in the building society industry, and in particular the range of diversification strategies being pursued by societies, illustrate that the industry is becoming much less homogeneous in nature. As a result only some societies will choose to be significant funders in housing markets, and even fewer will undertake a major development role.

However, these special housing markets will become important to the work of *some* societies in the 1990s, and their finance will have to be harnessed by their enabling, developing and managing partners if the crucial housing issues are to be addressed. The building society role in this partnership is far from passive. Societies have a unique perspective on housing given their experience across a range of markets together with their political and social status. They are a crucial partner with the potential to become an important voice in housing policy debate. Although all parties constantly highlight the importance of partnership, I am far from convinced that there is a real desire by the majority of providers and managers of housing to make such schemes central to their strategies rather than peripheral or supplemental. Without true partnership with building societies, social housing enablers and providers will continue to be starved of financial resources. The scale and structure of the funding required to tackle the real issues of housing supply and condition are not going to come from anywhere else.

REFERENCES

Boleat, M. (1990) *The Building Society Industry*, London: Allen & Unwin.
Building Societies Gazette (1987) 'Urban Renewal', Franey & Co., pp. 48–54.
Coleman, D. (1988) *Housing Policy: Unfinished Business*, London: Bow Publications, pp. 21–5.
Joseph Rowntree Foundation (1991) *Housing Research Findings*, no. 29, March.
Kleinwort Grieveson Securities (May 1988) *Building Societies*, Tunbridge Wells: Kleinwort Grieveson Securities.
Mortgage Finance Gazette (1990) Franey & Co., November, pp.75–9.
Slaughter, J. (1989), 'Girobank's Friendly Alliance', *The Observer*, 25 June, p.56.
UBS Phillips and Drew (1991) *Banks and Building Societies – Bloodbath in the High Street*, February.

4 The social and economic consequences of the growth of home ownership

Stuart Lowe

This chapter focuses on one aspect of the recent experience of home ownership: the rapid increase in the rate of capital accumulation in private housing and the closely related surge, during the mid-1980s, in the scale of equity withdrawal from the housing market. There are a variety of sources of such financial 'leakage', primarily from housing inheritances and from the proceeds of moving house, and the findings of a recent study will be used to indicate the scale of resources involved and how leakage occurs (Lowe and Watson, 1989). Such is the magnitude of the leakage that important social and macro-economic consequences are beginning to be generated. Owner-occupation has become for millions of households a significant generator of cash and wealth. Moreover, this financial gain has, over the past five to ten years, gradually become constituted independently from the occupational class structure. The consequences raise some important issues about the influence of housing on the nature and structure of British society in the late twentieth century and beyond.

It should be made clear, however, that although the scale of resources involved is very substantial indeed, this is *not* a recipe for a financial bonanza for all home-owners, as some authors have claimed. Saunders and Harris, for example, argue that:

> home ownership has come to represent the equivalent of a certificate of entitlement to share in the fruits of economic growth. It is, despite all the critics' claims to the contrary, quite literally a stake in the capitalist system.
>
> (1988, p. 10)

There *is* evidence of some remarkable developments, but home ownership is not a homogeneous housing tenure and does not bestow its benefits evenly. Several studies of low-income and marginal home-owners have pointed to the precarious hold of some households on the property market (Karn *et al.*, 1985; Thorns, 1981). Indeed, many authors cast considerable doubt on the capital accumulation potential of home ownership and conclude that if it does occur the tendency will be to intensify existing social divisions. Thorns argues that:

> the housing market functions to create a growing differentiation amongst owner-occupiers. . . . The evidence further suggests that the process of accumulation transfers wealth to those who already have substantial assets, thus reinforcing rather than reducing existing social inequalities.
>
> (Thorns, 1981, p. 28)

In a similar vein, Forrest concludes that ' . . . the processes currently underway in the British housing market . . . will amplify income and class differentials in housing' (1983, p. 213). Despite the scepticism of Thorns, Forrest and others it is now a reasonably secure argument that owner-occupation is a source of capital gain, and the associated leakage from this equity has grown dramatically in recent years. A middle mass of home-owners has benefited considerably. But this conclusion has to be read in the wider context of the restructuring of welfare and public-service provision in the 1980s, the impact of equity withdrawal on the management of the macro-economy, and the complex structure of owner-occupation as a housing tenure. In this sense, the extent of the redistribution of wealth, implied by Saunders and Harris (1988), is illusory because households have been required to defend and readjust their living standards in an era of privatisation of services, access to which is dependent on cash. The crucial connection in this game plan is that in the past five to ten years (and arguably before), owner-occupation has become an independent financial resource, and the trajectory of this autonomy is increasing. Questions about whether the gains from home ownership exaggerate existing social divisions, as has been claimed, or about their capacity to produce new social schisms separate from the traditional class structure, must take into account the wider role played by home ownership as it matures as the dominant housing form.

The growth of home ownership as a source of capital accumulation will be reviewed in this chapter, and particular attention will be paid to the closely related issue of the scale and sources of equity withdrawal. After this the chapter examines the implications of these developments in a policy context and within the social structure.

The chapter begins with a short profile of owner-occupation and an outline of some of the key issues which have confronted this tenure due to its expansion during the 1980s. This material provides an essential context within which to read the evidence of equity withdrawal and provides a cautionary note for the more extravagant claims from whichever 'side' of the debate. Home ownership is a very heterogeneous housing tenure, not least because over two-thirds of the population live in owner-occupied dwellings. In one sense, therefore, it is simply a mirror of society as a whole.

HOME OWNERSHIP IN THE 1980s – A PROFILE

In aggregate terms, owner-occupation currently accounts for about 65 per cent of households and is still growing at the rate of about one per cent per annum at the expense of the rental tenures. However, the current profile of the tenure is one of considerable unevenness and variation. The most important of these factors are in the distribution of home ownership by social class, by age of the heads of household, and the much publicised inter-regional disparities in house prices.

The relationship between social class and home ownership will be discussed later in the chapter. Here it is simply noted that virtually all the higher socio-economic groups are owners of at least one home, and if they are not it is commonly by choice. The most significant division is between skilled and unskilled manual workers, with about two-thirds of the former but less than 40 per cent of the latter being home-owners. There are also wide variations in the owner-occupancy rate according to age and very considerable differences in the rate of increase of owner-occupation in different age-groups. Among the 30–60 year olds the owner occupancy rate is currently about 75 per cent and grew rapidly, from about 50 per cent in the early 1970s. Among the older age cohorts the rates of change have been very much lower and at the moment owner-occupation accounts for less than 50 per cent of the over-60s, and declines with increasing age. This variation across the age structure will have important long-term effects, particularly on the growth of housing-based inheritances.

Home ownership is not evenly spread geographically or by price (both *between* regions and *within* them), and these factors have very significant repercussions on how to interpret the social and economic impact of mass home ownership. For example, over three-quarters of households in the southeast of England are owner-occupiers, but the ownership rate for the region covered by Cumbria and the northeast is only 50 per cent, and in Scotland home ownership remains a minority tenure at about 40 per cent of households.

The much-publicised disparity in house prices between the 'south' and the 'north' has important economic and social repercussions, and became significant, in the early 1960s, at a time when house prices began to increase above the general level of inflation. During this period prices in the northern regions have consistently been at about 75 per cent of national average prices, whereas London and the southeast have averaged between 25 and 50 per cent higher depending on the point in time of the house price cycle. This is a stable pattern and the rank order of all the regions in relation to national average prices has not changed over the past twenty years.

A house price cycle was also associated with the acceleration in house prices and has an inter-regional dimension. In the cycle prices rise for several years before the rate of increase slows and sometimes, as in the current phase of the cycle, declines in real terms. The pattern is for London and to a lesser extent the southeast to drive the market, with a price 'ripple' across the rest of the country. In every case since the house price cycle became significant the northern markets have adjusted relatively quickly by continuing to inflate when the southern market has turned down.

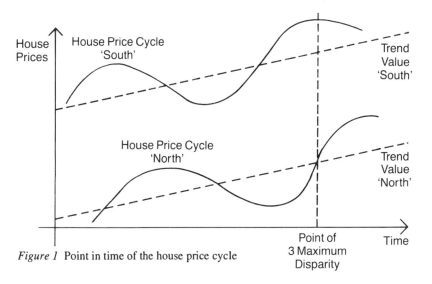

Figure 1 Point in time of the house price cycle

Figure 1 illustrates an idealised history of the house price cycle in England, showing the lag in the northern markets and the inter-regional disparity in prices. In estimating the capital gains and potential for equity leakage it is important to be sensitive to both the house price gap and the trend value of the price cycle. A number of studies have also pointed to the quite large disparities in house prices *within* regions, and this undoubtedly complicates the house price 'map'.

These issues raise important questions about the 'affordability' of home ownership and the relationship of the private housing market to the provision of social rented housing. Bramley has argued, for example, that housing subsidies for the development of new public housing should be concentrated in the southeast of England because of the problems facing first-time buyer entrants to the inflated housing market and the low level of council house re-lets in the south compared to the north (Bramley, 1989). It is not at all clear, however, that Bramley's conclusion is valid, partly because his analysis is not sensitive to the point in time of his estimates in relation to the house price cycle (he measured the north/south gap at its point of maximum disparity), and partly because of counterveiling evidence concerning the lower levels of income and higher unemployment, and the somewhat poorer quality of the housing stock in the north (Barnett *et al.*, 1989). In addition, the higher capital value of the southern housing stock creates a paradoxical situation in which a small flat in London can be exchanged for a substantial dwelling in other parts of the country, especially in the north. A property with a low accommodation value has a much higher monetary value in the south than in the northern regions.

Thus as owner-occupation has come to dominate the housing system major issues of access and affordability have also opened up. These issues include problems of inter-regional labour mobility (Munroe, 1988); homelessness, particularly in London (Bramley *et al.*, 1988); and housing the victims of relationship breakdown (Hughes, 1987). In this context it is important not to treat owner-occupation as totally separated from the rental sectors, particularly socially rented housing. In an era in which the financing of council housing and housing association accommodation is increasingly dependent on private finance and market rents, there are important inter-dependencies between owner-occupation and rental forms of housing.

Home ownership is clearly not a homogeneous tenure but is often spoken of as though it is a single entity. The evaluation of the evidence on home equity and leakage needs to avoid such a monochromatic picture. This applies also to the emotive treatment it has received in some of the social science literature, particularly on the issue of capital accumulation. Forrest concludes his paper by saying, 'Politically it is essential . . . to attack the claims made of home ownership as the route to affluence and a more equitable distribution of social and economic power' (Forrest, 1983, p. 215). Saunders and Harris, on the other hand, as we have seen, speak of home ownership as a 'certificate of entitlement' and a 'stake in the capitalist system' (1988). These stances cannot both be right, and they demonstrate the ease with which social science can become rhetoric and something of the power of home ownership to stir the emotions.

The next section of the chapter explains briefly the logic of the claim for rapid capital accumulation, and is followed by an account of estimates that have been made of the scale of equity withdrawal from the housing market over the past five years (Lowe and Watson, 1989).

CAPITAL ACCUMULATION AND EQUITY WITHDRAWAL –
THE EVIDENCE

Home ownership has been a real source of capital gains for about twenty years. This is because dwellings have risen consistently faster than the Retail Price Index (RPI) over that period and have out-performed other types of investment, and because the long-term costs of access to an owner-occupied dwelling are probably, on average, lower than that for renters. Moreover, owner-occupiers in the manual-worker social strata may have made gains at a similar rate as non-manual households (Saunders and Harris, 1988).

Evidence from the two main house prices series, those of the Halifax and the Nationwide Anglia building societies, indicate that housing has inflated at roughly twice the rate of the RPI in the past two decades. The reasons for this inflationary impetus in the housing market are complex. It is a consequence of a number of factors such as the demand and supply within the market, growing real incomes, and interest rate instability from about the early 1960s (Holmans, 1986b). It is also well known that very large subsidies have been pumped into owner-occupation, mainly via the tax system. Thus the abolition of the schedule 'A' taxation of owner-occupiers in 1963 (which was the tax on the imputed rental income enjoyed by owner-occupiers) in the context of the continuation of mortgage interest tax relief effectively created an incentive to home owners to use their house as an an investment. Owners receive what is in effect a subsidy on the access payments but pay no tax on the the imputed rental income *nor* on the capital gains made when they sell. It is in this context, from the mid-1960s, that house prices began to rise above the RPI, due to the financial incentives within the owner-occupied market and the capitalisation of these subsidies into house prices.

At the same time the key relationship between the earnings index and the house prices has remained, as a trend across the house price cycle, relatively stable; average house prices are normally at between three to four times average earnings. Housing is thus uniquely placed in the income and consumption ratios.

The second key to the accumulation potential of housing is its performance, relative to other physical and financial assets, as an investment. Over the past five years the value of private houses has risen on average by nearly 170 per cent (according to the Halifax Building Society prices index), whereas the value of equities listed in the FT-SE 100 share index peaked at about 140 per cent before the stock market collapse in the autumn of 1987 and is currently at about 120 per cent of the 1983 level.

This would not be a 'real' gain if the costs associated with mortgage

repayments and repair and maintenance exceeded the costs of renting
accommodation. In fact the average costs of owning and renting can be
treated, at the very least, as a neutral element in the financial equation.
Indeed, the evidence suggests that on average the costs associated with
renting are, in the long run, more expensive than those of home ownership.
Saunders and Harris point out that ' . . . it is legitimate to exclude mortgage
interest and routine maintenance from calculations of net gains to home
owners . . . as these costs may be equated with the rent which owners would
otherwise have to pay if they chose to put their money into some other form
of investment' (Saunders and Harris, 1988, p. 3).

But the crucial difference between housing and other forms of
investment is that, for the great mass of owner-occupiers, access to the
purchase of their home is normally through mortgage finance. Currently
about 80 per cent of the purchase price for first-time buyers, and about 60
per cent for movers within the market, is sponsored by mortgage finance.
Personal liability is thus limited by the use of institutional finance and
allows the buyers to 'gear up' their investments; they take the gains from
the whole investment less only the amount necessary to redeem the existing
mortgage. This process of setting and periodically re-establishing the
household's gearing ratio (the relationship between outstanding mortgage
debt and the current value of the house) is the key to the accumulation
potential in home ownership and is the reason why Pahl could claim, 'A
family may gain more from the housing market in a few years than would
be possible in savings from a lifetime of earnings' (1975, p. 291). A
calculation based on the past five years in the housing market illustrates this
point. A £40,000 house bought with a £10,000 deposit in London in 1983
had risen in value by 1989 to over £100,000, producing a profit in excess of
£40,000 even after deducting the mortgage and transaction costs.

These potential investment gains are not, however, evenly distributed.
The discussion in the previous section drew attention to the likely impact of
both the point in time of entry to the market and the place of entry, and
direction of subsequent moves, in the rate and size of capital appreciation.
These issues undoubtedly complicate the capital accumulation 'map', and
there is still a great deal of work to be done in disaggregating these gains
within both the social structure and the regional housing markets. But the
financial gain to the steadily swelling ranks of home-owners over the past
twenty years or so can hardly be disputed.

Equity withdrawal from the housing market

It has been suggested that even if there is capital accumulation taking place
in the owner-occupied housing market, these gains are only of 'paper' value

because they cannot be realised. Kemeny argues, for example, that ' . . . the capital gains made in owner-occupied housing do not generally accrue to anyone: they are simply passed from one owner to another' (1981). Ball also casts doubt on the accessibility of the capital accumulated in the owner-occupied housing stock by arguing in effect that they are not realisable gains (1983). In fact, this has not been the case during the past ten years or so.

So-called equity withdrawal occurs in a number of different circumstances and is defined as the replacement of some or all of the value of the dwelling by new lending. This occurs in two main ways; through housing-based inheritances (when the house of a dead owner is passed to relatives and then sold) or as the result of a number of different types of moving household. Here the 'leakage' occurs when the household increases its gearing during the move from one house to another (by taking a bigger mortgage for the purchase of the new house and keeping some of the proceeds of the sale of the previous house), or by 'trading-down' (moving to a less-expensive house).

Attention has tended to focus on the social consequences of housing inheritances, and there is now a substantial body of literature on this issue (Saunders and Harris, 1988; Munroe, 1988; Forrest and Murie, 1989; Harmer and Hamnett, 1990; Hamnett *et al.*, 1989). The source of the surge in equity withdrawal that began in the early 1980s was primarily due to moving households, however. The number of owner-occupiers who moved house increased by over 60 per cent between 1982 and 1987. It seems likely that this mobility was stimulated by relatively low interest rates and the increasingly competitive market for mortgage finance, following the intervention of the banks and the growth of a range of new lenders, including overseas banks.

The consequences of this growth in mortgage finance were discussed by the Bank of England (Bank of England, 1985) and in the papers of a number of City finance companies. The Bank of England's estimates of the scale of equity withdrawal were widely quoted, for example, in the *Inquiry into British Housing* chaired by the Duke of Edinburgh (NFHA, 1985), and in the Church of England's report *Faith in the City* (1986). These estimates, however, were not entirely satisfactory, and A.E. Holmans, Chief Economic Adviser at the Department of the Environment, made a new set of estimates based on an analysis of the flow of funds within the housing market. This report has been widely interpreted as indicating that inheritance is the primary source of equity leakage, and Holmans' estimates show that the massive sum of about £3,000 million per annum was taken out of the housing market through inheritances in the years between 1982 and 1984. But movers with and without a mortgage accounted for somewhat more than this in Holmans' estimates. In addition, there are a range of other types of movers, such as existing owner-occupiers who marry and sell one of the dwellings, or emigrants, or elderly people moving into residential homes. In this broad sense movers account, according to Holmans' figures, for about two-thirds of his narrow definition of equity withdrawal.

This section of the chapter describes the findings of a more recent study which has made estimates of the amounts and sources of equity withdrawal over the past five years (Lowe and Watson, 1989). It should be said at the outset that these estimates, and Holmans' calculations, provide figures of about the right magnitude and are the best available given limitations of data, but have some 'weak spots' due to the unavailabiltiy of recent data on some of the key accounting identities in the calculations. In particular, there has been no survey of moving households since 1977, although the housing trailer to the 1984 Labour Force Survey (LFS) provided a useful update on the number of moves that took place in 1984 and whether or not the transaction involved mortgage finance. With these limitations Table 1 presents the findings of estimates for different catagories of equity withdrawal for 1984. This is considered a 'secure' year for the estimates because of the LFS data, but estimates have been worked forwards in some cases to illustrate the growth in the scale of leakage in more recent years.

Table 1 Equity withdrawal from the housing market in 1984

	(£1000s)	%
Elderly households dissolved:		
Due to death	3,965	27.5
Other reasons	1,375	9.5
Other 'last-time' sales	1,492	10.3
Movers with a mortgage	4,165	29.0
Movers without a mortgage	553	3.8
Sales by private landlords	1,335	9.3
Sales by public authorities	1,543	10.6
TOTAL	14,428	

Source: Lowe and Watson, 1989, p. 5. This data is adapted from work of Holmans, 1986a. The estimates for movers with and without mortgages are dependent on accounting identities derived from the 1977 NHDS (Recently Moving Households, OPCS, 1983).

Each of the main sources of leakage is discussed below with estimates of the recent scale of cash involved using 1984 as a base year.

Equity withdrawal from housing inheritances

The significance of housing-based inheritances is clearly seen in the table. In their research Hamnett, Harmer and Williams derived a direct figure, from Inland Revenue data, of 143,980 estates which contain residential property passing at death in 1982–83 (see Hamnett *et al.*, 1989). This figure accords with the estimate of 150,000 elderly owner-occupier households dissolved that was made in the Lowe/Watson study, calculated from the Mortality Statistics, Population Trends and adjusted for an owner-occupier rate among elderly people using the General Household Survey (only about 45 per cent of the over-60s were owner-occupiers in 1984). Both these figures accord with the estimate made by City financiers Morgan Grenfell of 155,000 elderly owner-occupier deaths in 1986 (Morgan Grenfell, 1987).

As was suggested above, the inheritance issue dominates the debate on equity withdrawal. The articles or papers have all pointed to the growing significance of inter-generational transfers of housing wealth, a sum of cash amounting to nearly £4,000 million in 1984. And this flow of funds will steadily increase over the next fifty years or so as the younger cohorts, in which owner-occupation rates are much higher, gradually die out. The evidence from these studies of housing inheritance is fascinating and begins to paint, for the first time, a picture of the social consequences of one of the major sources of equity withdrawal.

Munroe's research, for example, confirms that substantial sums of money are indeed passed within families as the result of an inter-generational transfer of housing wealth. Her findings show that this money typically passes to the same or the following generation. These are people already well established in their housing careers; such transfers result in ' . . . deeper wealth divisions in the longer term between those who own houses and those who do not' (Munroe, 1988). Forrest and Murie's focus is on the uneven pattern of distribution both by social class and geographical area, and they make the useful cautionary point that although wealth accumulated through housing inheritances is significant we need to beware of over-simplifying the processes involved in its effect on existing patterns of inequality: ' . . . wealth accumulation and inheritance through housing will be highly differentiated and may be more likely to accelerate rather than smooth out social divisions' (Forrest and Murie, 1989).

These studies, and the Hamnett, Harmer and Williams research, point to similar conclusions: that housing-based inheritances will increase over the next thirty years, and that the cash released will gradually spread to a wider social stratum but, at the moment, is heavily concentrated among the better-off existing home-owners, with a strong geographical bias towards London and southern England. Tenants, of course, are by definition excluded from this source of wealth and, as a result, a core of low-income, manual-worker households will become relatively worse off.

Table 2 shows the results of the estimates made in the Lowe/Watson study of the number and value of housing inheritances between 1982 and 1987.

Table 2 Quantity and value of housing inheritances 1982–87

	1982	*1983*	*1984*	*1985*	*1986*	*1987*
Inheritances (000s)	156	149	150	141	146	148
Average value (£)	19,474	21,946	25,506	25,780	30,082	33,291
Total value (£000,000)	2,988	2,998	3,965	3,562	4,304	5,545

Source: Lowe and Watson, 1989

The table shows small fluctuations in the number of elderly owner-occupier deaths, which on such a short time scale is only partly accounted for by variations in death rates and the owner-occupation rates. The main cause of the fluctuations is that the value of inheritances in any one year is sensitive to the house price cycle discussed above. Within the cycle there is likely to be a certain amount of lag because beneficiaries may not sell the house straight away, waiting for a time when prices appear to be buoyant. But the surge in house price inflation in 1986 and 1987 clearly shows through, producing a figure for equity withdrawal from housing inheritances of well over £5 billion in 1987. The long-term trend is, of course, inevitably upwards due to the demographic changes discussed above.

Equity withdrawal by movers with a mortgage

The figures in Table 3 are based on estimates of the scale of equity withdrawal by movers with and without loans. They are secure figures in expressing the magnitude of leakage during moves, but the category of movers without loans is derived from Holmans' estimates based on the sample drawn from the Movers' Survey (OPCS, 1983).

Table 3 Estimates of cash proceeds for movers 1982–88

	1982	1983	1984	1985	1986	1987	1988
Movers with loans (000s)	511	569	631	639	809	817	1,176
Cash proceeds for movers (£000,000):							
With mortgage	3,393	4,780	4,165	4,552	4,905	6,261	10,088
No mortgage	469	519	553	588	656	742	720

Source: Lowe and Watson, 1989

These findings confirm the very large magnitude of equity being extracted from housing market processes by moving households. The figures fluctuate more than the results for housing inheritances and clearly reflect an extraordinary growth in the number of moves that took place between 1986 and 1988. This may well be a result of the greater availability of mortgage finance due to the intervention of new lenders and the change in the rules governing the quantity of money building societies are themselves able to buy in the wholesale money markets. The number of moves more than doubled between 1982 and 1987, and the amount of equity withdrawal associated with moving trebled in that period. The amount of moving that occurs is, of course, susceptible to the state of the housing market and interest rates, and has slowed during 1989.

Movers who do not buy

The significance of movers who do not buy another house is seen in Table 1. The total of equity released from the proceeds of these 'last-time' sales by movers for 1984 was over £2.8 billion, and accounted for over 25 per cent of the total withdrawn on the 'narrow definition' of equity withdrawal, which excludes sales by landlords. There are a number of categories in this group: elderly households who move into residential care, people who marry or re-marry, and emigrants. Data on the numbers of people in these catagories can be derived from research reports and government statisitics. The use of average house prices gives an estimate of the quantities of cash involved in these house sales (with an adjustment to the value of the properties being sold by elderly households; information from the National Housing and Dwelling Survey showed that the average value of property owned by widows was 9 per cent less than the average for movers as a whole).

Equity withdrawal by private landlords

This category of equity withdrawal is included because most of the money derived from sales to sitting tenants or on vacant possession goes to individuals who happen to be landlords. The sums involved in the mid-1980s are very considerable (about £1.3 billion in 1984), but with private renting currently accounting for only about 6 or 7 per cent cent of households, it will certainly decline in significance as a source of equity withdrawal. We are at the end of an equity withdrawal of massive proportions as private landlords over the past eighty years have sold out.

Money accruing to local authorities from the sale of council houses does not enter the personal sector, and so is excluded from consideration in these estimates. Eventually these houses will be significant in the context of equity withdrawal by individual households, through inheritance.

Other types of equity withdrawal and associated borrowing

In addition to the forms of equity withdrawal described above, a number of closely related forms of leakage and secured lending based on the value of the properties have become of increasing significance. The growth in these forms of equity withdrawal and related borrowing derive from changes in the structure of the mortgage market in the 1980s. When the high street banks entered the mortgage market in 1981–82 (taking 40 per cent of new lending in that year), the building society cartel broke down. This is also the period when a variety of 'new lenders' began to make an impression on the market for mortgage finance. Targeted particularly at the up-market sector, insurance companies – operating through intermediaries such as The Mortgage Corporation – and overseas banks expanded the competitive range within the mortgage industry. The 1986 Building Societies Act also expanded the range of activities which building societies could engage in, and the amount of finance they are able to raise in the wholesale money markets has been increased. Their dependency on the flow of funds from individual investors is thus much less important than it used to be in earlier decades when mortgage 'queues' were common. Government disapproval of the new lenders and the new forms of lending has gradually dissipated.

Refinancing and secured loans are the major areas of innovation in the mortgage market of the past few years. At the end of 1987 personal-sector dwellings were valued at about £750 billion while outstanding mortgage debt was less than £180 billion, suggesting ample scope for future development in this area. There are two main catagories of financing involved here: 1) changes over time in the house price/mortgage advance ratio, and 2) forms of refinancing without moving house.

1 One effect of the new scale and type of lending practices was a form of leakage due to the growth in the ratio of advances to house price. This ratio is prone to cyclical fluctuations, but the trend value has been moving upwards in the 1980s. It occurs when first-time buyers and existing owner-occupier movers take larger loans in relation to the value of the property being bought than has been the case previously. This implies that purchasers put less of their own savings into the transaction and, therefore, retain them for other purposes. Congdon and Turnbull calculate that the spending power of first-time buyers in 1982 was increased by about £800 million compared to 1980 by the increase in the ratio of advances to house price (Congdon and Turnbull, 1983).

The increase in the proportion of mortgage advances may be a response to higher house prices rather than the ability to retain personal savings for other uses. But in a situation of an abundance of mortgage finance – 100 per cent mortgages being not uncommon in recent years – the relationship between savings and advances suggests a large-scale equity withdrawal is potentially available. There was certainly an acceleration in the ratio of mortgage advances to house prices after the intervention of the high street banks in 1981–82. The advance to price ratio increased from 46.1 per cent in 1980 to 58.5 per cent in 1984 for moving owner-occupiers, and from 73.8 per cent to 85.5 per cent for first-time buyers. This is on the order of a 7 per cent increase for movers and a 4 per cent increase for first-time buyers more than would have been expected before the market became more competitive. Comparing house prices and the advance ratio for 1981 with 1984 suggests that about £2 billion more was lent in 1984 than would have been the case if advances, prices and proceeds had held their 1981 identities.

2 The availability of increasing amounts of mortgage finance and house price inflation enables households to use their housing equity as a source of cash without moving. There are three main types of refinancing that need to be accounted for in this context: remortgaging (i.e. replacing an existing loan with a new one without moving), increasing an existing mortgage, and using the house as collateral for a loan. The sums of money involved are substantial, but there is very little direct evidence on which estimates can

be based, and the statistics from the banks and the new lenders are poor. In their study of housing inheritance, Hamnett, Harmer and Williams found that 8 per cent of their respondents had remortgaged or borrowed money using the house to secure a loan at some time (1989). But there was no disaggregation of the two types of borrowing or any indication of how much was borrowed. Saunders and Harris (1988) found that 16 per cent of their sample of owner-occupiers had remortgaged and 9 per cent had used the house as collateral against a loan. But there is a definitional problem with these figures. Saunders and Harris attribute most of the money (86 per cent) derived from remortgaging for use on home improvements or repair work, but it seems likely that there is a confusion here between 'topping up' an existing mortgage and full remortgaging.

Holmans calculates a figure of only £45 million for full remortgaging for 1984, implying only about 5,000 cases, but the average for the three years 1982–84 is £140 million per annum. There is an inherent problem in calculating the scale of this type of equity withdrawal because the motive for remortgaging may be connected not with the desire to release cash but with the desire to get better terms for the mortgage in the increasingly competitive mortgage market. The growth of remortgaging is, however, a potentially major source of equity withdrawal in the sense used here, but more empirical work needs to be done.

The use of top-up loans for house improvements is known to have been a semi-legal source of equity withdrawal when part of an approved loan was not used for the designated repair or improvement work. It caused some concern because this new lending was eligible for tax relief, if it did not take the mortgage above the eligible amount. But in the new competitive environment topping-up an existing mortgage to release cash for general consumption rather than adding to the equity of the dwelling (as in the case of improvement work) is likely to have increased.

Borrowing money and using the house as collateral for the loan is closely connected to these forms of financing, and in this wider definition it has certainly increased in the last few years. It is counted for this purpose as a financial leakage because the equity would be used in the case of a default on the outstanding loan, in much the same way as a mortgage redemption before full term. This also requires a detailed evaluation of building society and bank statistics to determine its scale and pattern of growth.

Summary of estimates for 1984

By bringing all the estimates together from all the sources of equity withdrawal and related forms of borrowing, the magnitude of this issue becomes apparent. The estimates for the year 1984 are taken here because they are the 'base' year figures in the Lowe/Watson study and so are less subject to error. It should be reiterated that these are only estimates and are subject to a number of methodological problems, particularly that the movers' estimates are not sensitive to 'traders-down'. The amount of cash entering the personal sector from all categories of equity withdrawal and related borrowing is as follows: inheritances released, about £4 billion; other elderly households dissolved, £1.4 billion; and movers, about £5 billion. In addition, about £2 billion was identified which remained in the savings of first-time buyers and moving owner-occupiers as a result of the increase in the advance to house price ratio; £1.3 billion went to private landlords selling their properties; and an as yet unspecified amount of money was taken by refinancing via remortgaging, borrowing against the home, and topping up loans. All these sources put the total of cash entering the personal sector directly from housing market processes at about £15 billion in 1984.

THE SOCIAL AND ECONOMIC CONSEQUENCES OF CAPITAL ACCUMULATION AND EQUITY WITHDRAWAL

The final section of the chapter discusses the implications of the growth in the value of equity held in the form of private dwellings and speculates about the significance of the scale of equity withdrawal on consumption patterns and the way in which households are able to defend or promote their welfare and living standards as a result of access to cash generated from their housing. Direct evidence is currently very limited, but reference to the consumption sector debate indicates some ways in which financial leakage from owner-occupation might underpin a range of new social schisms and life chances.

One obvious consequence of the growth in the value of private housing is to set up a social cleavage between home-owners and tenants. As was argued above, rent costs over a lifetime are higher than mortgage and maintenance costs for most home-owners, and because the current rent system, in both the public and the private rental sectors, is designed to reflect more closely local capital values, the financial disparity between owner-occupiers and tenants will intensify. But even if the relative costs of owning and renting are excluded from the equation or are taken to have a neutral effect, tenants are completely cut off from any capital gain. In this sense an important consumption-based cleavage has developed within the social structure and had been underpinned by the residualisation of council housing during the 1980s. The interpretation of this social division needs to be sensitive to the wider restructuring of welfare during this decade, the complex structure of the housing market, and the impact of the housing market processes.

Social class and owner-occupation

The wealth and financial gain which have accrued from home ownership have begun in recent years to represent a net addition to the income and assets of owner-occupied households which are unrelated to earnings. Moreover, it appears to be the case that owner-occupiers across the social class spectrum have all gained in relative terms. Saunders and Harris conclude from the evidence of their 'Three Towns' study that ' . . . working class people have secured high rates of return from home ownership and do not seem to have fared any worse than higher social classes' (Saunders and Harris, 1988, p. 15). The estimates for the scale of equity withdrawal suggest that as much as 2 per cent of personal disposable incomes are accounted for by equity withdrawn from housing (Lowe and Watson, 1989). It also seems increasingly likely that the capital gain and potential for equity withdrawal common to all home-owners derives from processes which are not inherently related to socal class. The surge, since the early 1980s, in the scale of equity withdrawal, primarily through moving households and a variety of forms of borrowing related to housing equity, is an endogenous part of the housing market. Moving house and new borrowing grew rapidly in the mid-1980s and was strongly influenced by the increasing competitiveness between the institutional lenders. To the extent that the capital gains of the past twenty years or so and the associated leakage of cash in the past five to ten years have come to be independently constituted, a significant consumption division (between home owners and tenants) which overlies the orthodox class system has developed. It is not the case anymore that these processes are, in the words of Preteceille, ' . . . mere reflections of class situation' (Preteceille, 1986).

This does not mean that working-class households have achieved social mobility via the housing market, or that they systematically trade against their equity. As Murie and Forrests suggest from a study in Bristol, ' . . . There is little evidence in the statement of home-owners or in the circumstances surrounding their housing moves to support a view that home ownership is used to pursue investment or speculative entrepreneurial activity' (Forrest and Murie, 1989). What emerges most strongly from this study is the differentiation and variation in the private housing market over time and between places. Mobility in the housing market is closely correlated to occupational mobility, with different sections of the housing market serving quite distinct 'nationally' mobile careers, against more locally based occupations (Murie, 1989). The Movers' Survey (OPCS,

1983) and the Nationwide Anglia database show, however, that most moves are short distances, suggesting that the motive for moving, apart from job changes, may well be associated with needs and ambitions in the housing market. The evidence about people's motives and ambitions for moving house is very limited and a great deal more needs to be known about the sociology of the housing market and how moving behaviour feeds into the wider economy through equity withdrawal or investment.

The evidence for how equity withdrawn from the housing market is used in household budgets and savings is also very fragmentary. It seems likely, other things being equal, that leakage from housing does feed into consumer expenditure due to its impact on the savings ratio. Throughout the 1980s the personal sector's savings ratio fell, and it seems likely that equity withdrawal has contributed significantly to this trend (Lowe and Watson, 1989). The growth of equity withdrawal through housing inheritances, for example, puts a long-term downward pressure on the personal sector's savings ratio. This downward pressure will continue so long as owner-occupation continues to grow and the owner-ocupation rate among the elderly age-groups increases. When this pattern stabilises, as seems likely towards the middle of the next century, the pressure will cease, but the personal sector's savings ratio will have stabilised at a permanently lower level.

It is also likely that there will be demographic influences within the long-term trend of equity withdrawal. Households at different stages in their life-cycle and housing careers will respond differently with respect to the holding of debt and equity in their house during a move (or remortgaging without moving). Such restructuring of the debt/equity position is likely to occur not only when a household trades up or down, but also when the move is for other reasons and no significant change in house type is made. Trading-down, say on retirement, from an area of high house prices to a cheaper area and a smaller house will improve post-retirement income. Or moving from 'south' to 'north' during a change of job offers considerable scope for choices about improving housing standards, attaining a desired level of mortgage debt, or gearing up and extracting the maximum level of equity withdrawal. Consumption and living standards clearly are influenced by decisions made by householders during their housing careers. If the savings ratio is permanently changed by equity withdrawn from the housing market, then these issues clearly have an important bearing on the state of the macro-economy and how it is to be predicted.

There clearly is a danger of seeking to read too much into the effects of capital accumulation in housing. Murie and Forrest's cautionary findings in their Bristol study are significant. However, the most frequently expressed opinion on this issue, that the consequence of home equity accumulation is to underline existing class-based differentials in living standards, is becoming less sustainable. By the same token, Saunders' claim that owner-occupation, in conjunction with the privatisation of services, has lead to the creation of non-class cleavages in the social structure – ' . . . major fault-lines . . . which may come to outweigh class alignments' (Saunders, 1984) – is also unsustainable by evidence. By broadening the debate about home ownership to the possibility of accessing other services Saunders has, however, made an important step in evaluating the wider consequences of owner-occupation.

One of the problems with Saunders' 1984 position, and with the consumption sector theorists in general, is that there is no necessary congruence between, on the one hand, the division of services into public and private sectors and, on the other hand, the emergence of major cross-cutting social divisions (Franklin and Page, 1984; Taylor-Gooby, 1986). There is no explanation of the processes which link owner-occupation to access to other services, and so to the creation of independently constituted 'sectoral cleavages'. There are, however, important inter-dependencies between home ownership and access to private forms of service provision and welfare choices, which are potentially linked by the scale of cash leakage from the housing market. It is here that the evidence of large-scale cash withdrawal from the housing market may be of significance. This is because the acceleration of private forms of provision – in health care, care of the elderly, leisure and the arts, transport, education – are contingent on an effective demand, on the ability of consumers to finance the costs of access, and on the market's willingness to adjust to these demands and opportunities.

For example, in the past ten years private-sector sheltered housing has burgeoned into a large and sophisticated market. There are probably about 55,000 units of this type of accommodation, the majority of which has been built in the past five years. From the initial ideas of the developers MaCarthy and Stone for cheap retirement homes, there is now a very diverse range of products – starter retirement homes for 'empty nesters', second-stage homes, homes for the frail elderly, up-market retirement villages. All these are non-subsidised developments which clients buy on the open market, in most cases using their housing equity. A number of studies in the north of England show that the entrants to private sheltered housing are drawn from home-owners of quite modest means and position in the housing market (Williams, 1988; Oldham, 1990).

On the supply side, the growth of the retirement homes industry is closely associated with the search of the building industry for new and lucrative markets, using public-sector provision as its model. The relationship between equity and the development process can clearly be seen in the much greater density of private sheltered housing in the south of England, although as house prices in the north climb the market will undoubtedly ripple out from its southern stronghold. Thus there are trade-offs available between consumption and housing standards which are not inherently related to social class but are more responsive to the supply and demand equation for particular markets of provision. The growth of private health care, the ownership of cars, private tuition for children, private domiciliary services, private residential care of the elderly – all are examples of marketed services which in part can be sponsored directly or indirectly from this source.

Moving provides an early opportunity, and probably the first of several chances, for owner-occupiers to trade against their housing equity for commodities or services. It also seems likely that lending within the family occurs, and probably on quite a large scale, so that the inter-generational transfer of resources will be taking place not only on the death of an elderly home-owner but at various stages in family life-cycles and housing careers. Thus it is important to note that different types of equity leakage may be used differently and have different effects on the social structure. It does not follow, for example, that leakage from movers will be reinforcing existing inequalities in the way suggested for housing inheritances by Munro (1988) and by Forrest and Murie (1989).

In the context of housing-based inheritances Hamnett, Harmer and Williams (1989) show that nearly half the financial receipts by the beneficiaries in their sample were invested in financial assets, mostly deposited in building societies. About one-quarter was spent on general consumption (which indicates an addition to general consumption in 1984 of about £1 billion), and a further 27 per cent was used on buying a first or second home, or on home improvements. These are very useful indicators as far as inheritance is concerned but, as the authors note, 'What we cannot predict from the results is how individuals will use these financial assets in the longer term' (Hamnett *et al.*, 1989). This is particularly the case with such a high proportion deposited in short-term investments and with no evidence of how the addition of this money affected existing spending patterns and the household budget as a whole.

We know also that a very high proportion of moving elderly home-owners of households that were dissolved entered private residential homes, presumably reserving the proceeds of the sale to pay the charges. There is limited evidence from the OPCS Movers' Survey that moving owner-occupiers use a proportion of the proceeds on refurbishing and improving the new house. People borrowing against the house or taking cash from remortgaging are almost by definition likely to be doing so for consumption rather than saving. Thus different forms of equity leakage may have a different social impact and are very likely to follow different patterns of use.

If it is the case that the expansion of privately marketed forms of provision during the 1980s has been sponsored in part from the huge amounts of 'non-class' cash derived from housing, there are important implications for the nature of British society, particularly for the restructuring of welfare opportunities. Millions of households, in the face of declining and less effective public provision, have to some extent been able to defend their welfare standards and extend their access to services through this means. This is not to argue that people necessarily have a preference for private services, let alone an innate, genetically determined desire for individual and private solutions (Saunders, 1990). This may or may not be the case, but in a policy context users of services have to make decisions about their welfare standards and how best to defend or satisfy them. This argument fits with the notion of 'coerced exchanges' advanced by Mishan in the context of transport policy (1967) and by Dunleavy (1980), that consumers may, in effect, be compelled to opt for private solutions as a result of the changing standards and role of publicly provided services.

Privatisation of services does not necessarily extend choices or social advantages or improve standards. It may do some or all of these. The point here is that households may have to switch to private solutions at a certain point in time. In this context it is the ability to move *flexibly* between the public and private sectors of provision that is partly sustained or improved by home equity (Lowe, 1986). Health, transport, education and the other services do not generate an impact on the social structure or access to services in this way, because they are not (directly, at any rate) sources of wealth. But insofar as housing can inject cash into household budgets it is potentially, at the very least, able to extend welfare choices and sustain the expansion of the supply of private services.

It seems likely that the potential of home-owners to convert their own housing equity (and housing-based inheritances) to social advantages will accelerate during the 1990s and beyond. The rate of capital gains made in housing is closely allied to the point in time of entrance and length of time in the market, so it can be supposed (assuming there is no catastrophic collapse of the housing market – which seems unlikely) that the spread of owner-occupation of recent years enables relatively new entrants to gear-up in later years. Housing inheritances will become a common feature of the middle years of a considerable majority of the population. How these resources are used or saved is an open question, but the scale of resources involved is massive and as time goes by the break between housing-generated consumption and social class is likely to widen and stabilise.

As far as the 1980s are concerned, it seems increasingly likely that the scale of resources involved in the process of capital accumulation and financial leakage from the private housing market may have affected the shape, and the pace, of the restructuring of welfare and public services away from the public domain and towards individualised modes of service use and privately marketed forms of provision. Given that these possibilities are unavailable to tenants, home ownership has become an autonomous influence on the social system, and in effect an independently constituted consumption cleavage. Living standards and welfare choices for the middle mass of home-owners became during the 1980s closely connected to their stakes in the housing market.

REFERENCES

Ball, M. (1983) *Housing Policy and Economic Power*, London: Methuen.

Bank of England (1985) 'The housing finance market: recent growth in perspective', *Bank of England Quarterly Bulletin*, March.

Barnett, R., Bradshaw, J. and Lowe, S. (1989) *Not Meeting Housing Needs*, Preston: Northern Group of Housing Associations.

Bramley, G. (1989) *Meeting Housing Needs*, London: The Association of District Councils.

Bramley, G. *et al.* (1988) *Homelessness and the London Housing Market*, Occasional Paper 32, Bristol: School for Advanced Urban Studies.

Church of England (1986) *Faith in the City*, London: Church House Publishing.

Congdon, T. and Turnbull, P. (1983) *Can the Upturn in Consumer Spending Continue?*, London: L. Messel & Co.

Dunleavy, P. (1980) *Urban Political Analysis*, London: Macmillan.

Forrest, R. (1983) 'The Meaning of Homeownership' *Space and Society*, vol. 1, pp. 205–16.

Forrest, R. and Murie, M. (1989) 'Differential Accumulation: wealth, inheritance and housing policy reconsidered', *Policy and Politics*, vol. 17, no. 1, pp. 25–39.

Franklin, M. and Page, E., (1984) 'A Critique of the Consumption Cleavage Approach in British Voting Studies', in *Political Studies*, vol. 32, pp. 521–36.

Hamnett, C., Harmer, M. and Williams, P. (1989) *Housing Inheritance and Wealth: A Pilot Study*, report to the ESRC and Housing Research Foundation.

Harmer, M. and Hamnett, C. (1990) 'Regional Variations in Housing Inheritances in Britain', *Area*, vol. 22, no. 1, pp. 1–15.

Holmans, A.E. (1986a) 'Flows of Funds Associated with Purchase for Owner-Occupation in the United Kingdom 1977–1984 And Equity Withdrawal from House Purchase Finance', Government Economic Service Division Working Paper No. 92, London: Departments of the Environment and Transport.

Holmans, A.E. (1986b) *Housing Policy in Britian*, London: Croom Helm.

Hughes, D. (1987) *Housing and Relationship Breakdown*, London: National Housing and Town Planning Council.

Karn, V., Kemeny, J. and Williams, P. (1985) *Home Ownership in the Inner City*, Aldershot: Gower.

Kemeny, J. (1981) *The Myth of Home Ownership*, London: Routledge and Kegan Paul.

Lowe, S.G. (1986) *The City After Castells*, London: Macmillan.

——(1987) 'New Patterns of Wealth: The Growth of Owner-occupation', in Walker, R. and Parker, G.(eds), *Money Matters*, London: Sage.

Lowe, S. and Watson, S. (1989) *Equity Withdrawal From the Housing Market: A Reappraisal*, report to the Joseph Rowntree Memorial Trust, York: University of York.

Mishan, E.J. (1967) *The Costs of Economic Growth*, Harmondsworth: Penguin.

Morgan Grenfell Economic Review (1987) *Housing Inheritance and Wealth*, London: Morgan Grenfell Economic Review, no. 45, November.

Munroe, M. (1988) 'Housing Inheritance and Wealth', *Journal of Social Policy*, vol. 17, no. 4, pp. 417–36.

Murie, A. (1989) *Supply, Transition and Differentiation: Perspectives on Home Ownership*, Bristol: School for Advanced Urban Studies (mimeo).

National Federation of Housing Associations (NFHA) (1985) *Inquiry into British Housing*, London: NFHA.

Oldham, C. (1990) *Moving in Old Age: New Directions in Housing Policies*, London: HMSO.

OPCS (1983) *Recently Moving Households*, London: HMSO.

Pahl, R. (1975) *Whose City?* 2nd edition, Harmondsworth: Penguin.

Preteceille, E. (1986) 'Collective Consumption, Urban Segregation and Social Classes', *Society and Space*, vol. 4, pp. 145–54.

Saunders, P. (1984) 'Beyond Housing Classes', *International Journal of Urban and Regional Research*, vol. 8, pp. 202–27.

——(1990) *A Nation of Home Owners*, London, Unwin Hyman.

Saunders, P. and Harris, C. (1988) 'Home Ownership and Capital Gains', paper for the International Conference on Housing and Urban Innovation, Amsterdam.

Taylor-Gooby, P. (1986) 'Consumption Cleavages and Welfare Politics', *Political Studies*, vol. 34, pp. 592–606.

Thorns, D. (1981) 'The implications of differential rates of capital gain from owner occupation for the formation and development of housing classes', *International Journal of Urban and Regional Research*, vol. 5.

Williams, G. (1986) *Meeting the Needs of the Elderly – Private Initiatives or Public Responsibility*, Occasional Paper 17, Manchester: Dept of Town and Country Planning, University of Manchester.

——(1988) 'Private-sector Provision of Sheltered Housing; Meeting Needs or Reflecting Demand', Paper delivered to British Society of Gerontology, Swansea, University College, September 23–5.

5   Private rented housing and the impact of deregulation

A.D.H. Crook

INTRODUCTION: THE OBJECTIVES OF GOVERNMENT POLICY

Whereas the promotion of house ownership was a major preoccupation of the government between 1979 and 1987, the third Thatcher administration placed much more emphasis on rented housing than did the previous two. In doing so, it was pursuing its broad privatisation objectives and therefore trying to create a more plural and market-oriented independent rented sector, whilst restricting the role of local authorities to an 'enabling' one and transferring their existing stock to a range of non-statutory landlords like housing associations, co-operatives and trusts. The government wanted new provision to be made, not only by housing associations but also by commercial landlords. The government thus set the scene for a major restructuring of housing tenure, reversing trends that have been in place for many decades. This chapter tries to assess its deregulatory policies with respect to commercial private landlords and the private rented sector in general. It examines the reasoning behind the new policies and the lessons that need to be drawn from past policies, and it describes the deregulation and associated policies and the likely impact they will have.

Regardless of the several attempts by governments to halt it (most recently, until 1988, by measures taken in 1980), the private rented sector's decline has continued unabated over, at least, the past seventy years, albeit the pace of decline has varied. A recent projection showed the decline continuing, although at a reduced rate (Whitehead and Kleinman, 1986). From a total of 1,856,000 households in England in 1981, it was projected to fall to 1,667,000 in 1991 and 1,339,000 by 2001, a fall from 11 per cent of all households to 7 per cent by 2001. The projection was done on 1981-based demographic projections of different household types, combined with projections of the propensities of each of these types to be private renters.

Despite the fact that a number of contemporary economic and social trends might be thought to have increased demand for private renting during the 1980s (e.g., increased mobility, growth in divorce and separation, increased local authority rents), the most recent evidence suggests that the above projection was already too high by the mid-1980s. By the end of 1987 the percentage of dwellings privately rented had fallen to 8 per cent. Moreover, very little of this was generally available to meet the demand for rented housing by the young and other mobile groups. Of the 8 per cent, 2 per cent was rented from employers (and not therefore generally available to the public), 4 per cent were unfurnished long-term lettings (most being sold to owner-occupiers when they became vacant), leaving only 2 per cent let as ready-access furnished housing.

Whilst the government's deregulation policies are designed to combat this shortage of ready-access housing, it is important to recognise that the changed emphasis of policies reflects the growing interest by independent commentators as well as by governments, both in Britain and abroad, in reviving private investment in rented housing (NFHA, 1985; Kemp, 1988; Maclennan, 1988). The reasons include a growing concern about the shortage of rental housing in a period of fiscal austerity, the way the pervasive physical decay of private rented stock threatens to undermine inner-city neighbourhood renewal programmes, and the need to house workers moving from economically distressed peripheral regions to jobs in areas of economic growth. The lack of any new building for private renting in Britain on any scale since the 1930s means that there is little choice for anyone moving jobs and wanting to rent tolerable standard housing in desirable neighbourhoods. In addition, an important motive for the British government's deregulatory policies is its ideological preference for privatisation and its hostility to direct provision by elected local authorities.

In its 1987 White Paper it was not surprising therefore that the government argued that private renting was a good option for people who needed mobility and did not want to be tied to home ownership. It argued that rent restrictions reduced landlords' returns and gave them no incentive to stay in the market, whilst laws on security deterred people from letting temporarily. The supply of private rented housing had shrunk below what was needed and partial deregulation was proposed to stimulate supply, building on the more limited initiatives taken in 1980 (Crook, 1986; DoE 1987). Whilst the government saw the lifting of controls on landlords as the main mechanism for reviving investment, it is also being recognised (if only in part by the government) that such a revival requires a restructuring of housing subsidies to enable private renting to compete with other tenures.

The need for this is evident from a consideration of the reasons why private renting has declined in this century and the role it plays today. An understanding of these is essential to a reasoned analysis of the likely success of deregulation.

THE DECLINE OF PRIVATE RENTING AND ITS CURRENT ROLE AND PROBLEMS

There has been a major restructuring of housing tenure in Britain this century (Malpass and Murie, 1987). In the last one, capital markets were less developed. Landlords acted as intermediaries between housing producers and consumers. They borrowed the money needed directly from small savers to buy houses and then rented them out to households (who were drawn from all social classes) for a weekly rent. In this century building societies have largely replaced landlords, providing a more secure, tax-efficient and liquid repository for small investors' savings. Hence landlords have lost their traditional sources of finance and become dependent on societies if they want to raise funds. Consumers who can afford to repay mortgages (and raise any necessary deposit) are better off borrowing from a building society to buy their own house to live in instead of renting it privately (Nevitt, 1966). As a consequence, two-thirds of dwellings in Britain were owner-occupied by 1987, the steady growth in this proportion fuelled by tax concessions to owner-occupiers, inflation, deregulation of financial markets and the wider availability of 100 per cent mortgages.

Current policies to revive private renting are a return to earlier Conservative preoccupations with ending controls. However, it is the low level of effective demand for private renting, allied with political uncertainty, poor reputation and tax/subsidy policies which have been just as important as regulatory measures in explaining decline (Doling and Davies, 1984; Hamnett and Randolph, 1988; Harloe, 1985; HCEC, 1982; Holmans, 1987; Nevitt, 1966). Decline has been accompanied by the expansion of owner-occupation. Landlords cannot offer housing at rents which are competitive with house purchase costs for those who can afford to buy, while social rented housing has provided alternatives for those who cannot afford to buy. There has been a weak demand from those households who could afford to pay a rent giving landlords a competitive return (but for whom owner-occupation is more attractive), whilst the incomes of those who cannot afford to buy limits the rents they can pay. It is this low demand, unaccompanied in post-war years (unlike some other countries) by supply-side subsidies, which is the fundamental cause of the decline, not rent and security legislation. Subsidies were available in the 1920s, of course, for all newly built houses, including those intended for owner-occupation as well as for private renting.

Whilst there is some heterogeneity in the contemporary role of private renting, overall it has an over-representation of households outside the labour market and without children (Bovaird *et al.*, 1985; Whitehead and Kleinman, 1986; Todd *et al.*, 1982; Todd and Foxon, 1987).

The unfurnished sub-sector (4 per cent of all dwellings) houses mainly elderly households and others of longstanding residence, reflecting the fact that private renting was the majority tenure when many of them set up home. They have secure but often poor-quality accommodation. Most have protected statutory tenancies and their rents are registered fair rents. Whilst evidence shows that investors are prepared to acquire this stock, it is almost exclusively done for speculative purposes, often involving builders and property dealers, new to landlordship, buying up property with sitting tenants from long standing, often individual landlords and doing it up with improvement grants in the meantime so as to sell to owner-occupiers when the current tenants leave or die. Rates of return on vacant possession value from rents are low (typically 5 per cent gross, or 3 per cent net of management and maintenance) and non-speculative investments do not give landlords returns that are competitive with other investments with similar risk and liquidity characteristics. Landlords thus almost always sell rather than re-let vacancies (Crook and Martin, 1988; Crook, 1989a; Paley, 1978).

By complete contrast, the furnished sub-sector provides short-stay, ready-access housing with low transaction costs for young single and other mobile, non-family households. It is important to stress that private renting houses only a minority of younger households. In 1985 only 15 per cent of those aged under 30 were private renters and already, by that age, 51 per cent were owner-occupiers (OPCS, 1987). Moreover, in contradiction to conventional wisdom, very few job-movers use private renting and most new households (especially married-couple households) set up in other tenures. Physical conditions in this sub-sector are poor. To some extent this can be explained by the age of private rented stock (two-thirds were built before 1919), but not entirely (DoE, 1988). Some of the worst conditions are in houses in multiple occupation (HMOs). In a 1985 survey, 80 per cent were found to be defective on at least one of the grounds of poor management, inadequate amenities or over-occupation (Thomas with Hedges, 1986). Mostly these were shared with strangers, and poor conditions promoted social tensions among tenants.

Many tenants in the furnished sub-sector also had insecure tenures. Whilst most tenants were, up to 1988, in principle protected by statutory security of tenure, many lettings were put beyond the scope of Rent Act protection (and thus also fair rent registration) by landlords who deliberately used devices like letting on licence agreements rather than on tenancies or by sham use of genuine exclusions, such as holiday lettings.

Up to half of recent new lettings in the 1980s appear to have been made in this way (HCEC, 1982; Todd and Foxon, 1987). Moreover, almost all rents of tenancies were privately agreed rather than registered fair rents and were high in relation to tenants' incomes, as well as the quality of accommodation provided. Tenants in the worst housing tended not to complain about it, however. They neither understood the system nor felt confident to risk complaining, because they feared landlords would harass and evict them as a result, and when they knew there was a shortage of good quality housing they could afford.

Landlords in this furnished sub-sector tend to be smaller in scale than those in the unfurnished sub-sector and there is evidence of continued willingness to invest in this sub-sector. Landlords are willing to buy empty property at vacant possession prices to let out, and most subsequent vacancies are re-let (Crook, 1989a; Todd and Foxon, 1987). Moreover, furnished rents give higher returns on vacant possession price than unfurnished rents. A recent case study, for example, found returns of 14 per cent gross and 9 per cent net of costs, excluding capital appreciation (Crook, 1989a). In many ways it is possible to regard this sub-sector (which is the target of government policy) as *de facto* already deregulated before the *de jure* deregulation of new lettings in 1989. Landlords let in ways which enabled them to get vacant possession, and rents were not set by rent officers. Moreover, they made competitive returns. However, it is important to note that in order to make such returns from letting to predominantly low-income tenants, landlords have had to neglect property repairs and to let on insecure terms at rents which took up high proportions of tenants' gross incomes. This has two significant corollaries. First, letting in this way gives landlords, in general, an unsavoury reputation and enhances negative images of landlordism. Indeed it reinforces the image of 'Rachmanism' which has come to haunt private renting and threatens to undermine any attempt to get 'responsible' investment (on Rachman see Committee on Housing in Greater London, 1965). The second corollary is that, prior to 1989, this sort of investment was inherently risky. Demand depended on the income maintenance systems available to the unemployed and to students. In the mid-1980s councils were starting to refer unusually high rents to rent officers and to challenge the validity of licences when licensees claimed housing benefit. Costs depended on local authorities' willingness to enforce repair and other standards, and many authorities were becoming more active in this regard in the mid-1980s (Crook, 1989b). Finally, landlords had to let outside the Rent Acts to maintain rent and the liquidity of their investments. Up until 1989 this was a high-risk area and a risk premium was required if landlords were to remain in the sub-sector.

Whilst there are well-paid economically active households at the better end of private renting, the tenure has become increasingly occupied by economically inactive and low-paid non-family households. In 1985 median household incomes in the unfurnished and furnished sub-sectors were 48 and 80 per cent, respectively, of that for all households, reflecting the numbers who are economically inactive and the low pay of those in work (OPCS, 1987). Housing Benefit is therefore crucial in helping low-income tenants to pay rent. Moreover, whilst both social-rented and private tenants have become increasingly concentrated at the bottom of the income distribution over the past three decades (Bentham, 1986), the landlords of the former, unlike the latter, have had central government subsidy on their stock. Without such subsidies on the private rented stock, a considerable gap has opened up between the rents landlords need in order to give them a competitive return and what can be afforded by low-income tenants. All in all, without subsidies, landlords can compete with neither individually subsidised owner-occupation nor collectively subsidised social-rented housing in providing reasonable quality rented housing whilst also making competitive returns.

This is what a Parliamentary committee in the early 1980s referred to as the 'central dilemma' (HCEC, 1982). The then legal and financial framework met the needs of neither tenants nor landlords. For example, furnished tenants paid high rents for poor and insecure accommodation. The landlords either let within the Rent Act and accepted below-market returns or operated outside the legal framework. The dilemma was that rents would need to rise considerably to give landlords competitive returns on providing habitable accommodation but that these rents would not be affordable. There is no unambiguous evidence as to the rate of return needed to keep landlords in the sector and to attract new investment (HCEC, 1982; Price Waterhouse, 1989). Nevertheless, since landlords can expect real increases in rents and capital appreciation, it may be better to compare returns from renting with those from equities, rather than building societies or government bonds. Even so, the 3 per cent net of expenses being earned by many unfurnished landlords in 1982 was clearly inadequate in the face of the 9 per cent gross, 6 per cent net then advocated by the British Property Federation (HCEC, 1982), or the 12 per cent gross, 10 per cent net then favoured by the Small Landlords Association – although the latter did accept that required returns would be lower if landlords were guaranteed repossession on reasonable terms, thus making it feasible to expect capital gains (HCEC, 1982). Whilst many furnished landlords were making this level of return, it must be remembered that it was a highly risky investment and was not being made on accommodation of a good standard. More recent evidence suggests that, whilst smaller private individual investors may be willing to make such equity investments, there will be little forthcoming from the major financial institutions because of the political risk of restraint being placed on rent increases linked to indices like wages. Instead, it is more likely that investors will be looking for returns from rents to cover debt provision, with perhaps returns of 12 per cent net required under current conditions (Mallinson, 1989).

Whatever the exact figure, or its basis, even the minimum of the returns noted above implies a doubling of rents on existing property of good quality and perhaps a quadrupling on new building. Thus rents would have to rise considerably to give landlords a competitive return. It is unlikely that such rents could be paid by existing private tenants without hardship or considerable additional subsidy. It is therefore unlikely that deregulation *per se* will lead to market rents rising to the levels which would yield the competitive returns identified. In other words, such returns imply rents above the level the market would bear.

It is true, of course, that *de facto* 'market' rents agreed before 1989 had been struck in the 'shadow' of the Rent Act. That is to say, landlords took account of the threat that tenants could get a fair rent registered and therefore did not pitch the 'market' rent too high so as to reduce the risk of tenants approaching rent officers when they considered the rent excessive. It is also true that regulation may suppress demand by fixing rents below those which tenants would willingly pay. Nevertheless, where appropriate comparisons between fair and market rents can be made, the latter were between 20 and 30 per cent more than the former (HCEC, 1982; Todd *et al.*, 1982). Even if the shadow effect were substantial, this suggests that *de jure* market rents would not rise to the levels required to make returns competitive.

RELEVANT PROVISIONS OF THE HOUSING ACT 1988 AND FINANCE ACT 1988

The relevant provisions of the Housing Act came into force on 15 January 1989 (Arden, 1989). They divide private renting into two groups, depending on the date tenancies started. Existing tenancies are left broadly unchanged, but all new lettings made on or after the commencement date are at market rents, with provisions for some security. The Act thus instituted a process of 'creeping deregulation' since re-letting of existing tenancies becoming vacant will also be at market rents.

Existing protected tenants lose neither their security of tenure nor their right to get fair rents registered. Their other existing rights are both modified and strengthened in two important ways. First, rights to succession have been modified to speed up deregulation. Spouses continue to succeed as statutory tenants, but when other qualifying relatives inherit a tenancy, it becomes assured (this applies to all second successions). Second, the laws on harassment were strengthened. An offence now occurs if landlords act in ways likely to interfere with the peace and comfort of their tenants. This means that it is no longer necessary to prove intent on landlords' part when bringing actions for harassment against them. In addition, to act as a disincentive to harassment, tenants can claim exemplary damages in a civil action if they are illegally evicted, up to the amount of capital gains landlords make from subsequent sales of the vacated properties. Tenants so evicted can get legal aid to commence such actions. These changes are intended to reduce temptations landlords have to evict existing protected tenants in order to let in the deregulated sector or sell with vacant possession.

All new lettings (including re-lets) which would have been Rent Act lettings in the past are subject to the market. Landlords have a choice between two types of letting, each let at market rents but with different security arrangements. First, there is a modified form of the 1980 Housing Act assured tenancy for which it is lawful to charge a premium and where there are freely negotiated rents. Security is provided for, since expiring fixed-term assured tenancies can run on as periodic tenancies provided, of course, that tenants are prepared to pay the rent. There are also some restricted rights of succession. Security is subject to revised, simplified and extended mandatory and discretionary grounds for eviction such as, respectively, thirteen weeks' rent arrears and persistent delays in paying rent. So far as rents are concerned, primacy of contract is the determining factor. Rents which tenants agree to pay for a fixed-term tenancy cannot be changed other than in a manner spelt out in the agreement. External agencies cannot be asked to alter these rents. Whatever rents tenants have agreed to pay, must be paid. The Rent Assessment Committee can, however, be asked to adjudicate on what are appropriate increases in market rents, where statutory periodic tenancies have arisen and where assured periodic tenancy agreements do not specify the basis for determining rent increases, such as the movement in the retail price index. The previous 1980 Act requirements for landlords of assured tenancies to be approved by central government and for qualifying lettings to be newly built or improved have been dropped.

The second choice landlords have when making new lettings is to let assured shorthold tenancies. These are a modified version of the 1980 Act protected shortholds. There is no security for tenants beyond a fixed term, which can be as little as six months. During the first contractual shorthold an application can be made to the Rent Assessment Committee for a determination of the market rent, taking account of the limited shorthold security. This determination can only be made if the market rent agreed by landlord and tenant is considered excessive and if there is information available on the market rents of comparable assured tenancies. If a rent is determined, it becomes the maximum payable.

In providing these alternative ways of letting, the government was anticipating that rents for assured tenancies would be higher than those for assured shorthold tenancies. Tenants would, it was thought, pay more for the greater security of the former, thus compensating landlords for their lower liquidity.

In expecting that market rents would be higher than registered fair rents, the government pointed out that all tenants receiving Housing Benefit would get any increases funded entirely by extra benefit. This is because, under the revised system introduced in April 1988, the amount of Housing Benefit paid to claimants depends on their income and household circumstances rather than on the rent they pay (Kemp, 1987a; Ward and Zebedee, 1989). Provided their income and household circumstances remain unchanged, therefore, recipients' rent increases are fully reflected in additional Housing Benefit payment. In other words, their marginal costs of housing are zero and they have no financial incentives, therefore, to negotiate with landlords to keep rents down. This potentially puts significant costs on the Housing Benefit system. A more general problem, if a large proportion of tenants receive Housing Benefit, is that it is hard to see how market rents can be struck between willing tenants and landlords when so many of the tenants are likely to have their rents fully underwritten by Housing Benefit, thus substantially modifying demand. This problem will be more intractable if there is not an increase in supply, and therefore increased choice after deregulation.

To regulate (and restrict) the government's apparently open-ended commitment to pay for whatever rent and rent increases are agreed to by tenants and landlords, a number of administrative devices have been introduced, extending existing practice to take account of the deregulated market. (The existing incentive arrangements, limiting central government subsidy on rents above a threshold, remain in place for existing regulated tenancies with unregistered rents.) First, rent officers will be able to notify local authorities if they believe tenants occupy too much space or rent very expensive properties. In the latter case authorities' entitlement to subsidy is unaffected. In the former it will be limited. More generally, market rents (unless fixed by the Rent Assessment Committee) paid by all Housing Benefit claimants will be referred to rent officers for validation as reasonable for the purposes of the subsidy paid by the Department of Social Security to cover the costs of local authorities who, of course, administer the system and pay out benefits. The determinations rent officers make will be based on a rate of return method allowing for capital gains, where no direct evidence of competitive market rents is available (Price Waterhouse, 1989). The government will reimburse local authorities the full cost of paying benefit on reasonable rents, but if they choose to pay benefit on rents which are judged 'unreasonable', they must bear the full cost of the difference. There are financial incentives, therefore, for local authorities not to pay out full benefit on unreasonable rents. In these circumstances and where authorities use their discretion not to pay out in full on unreasonable rents, claimants will have to find the difference themselves – or get landlords to reduce the rent.

Two other pieces of legislation are relevant to a consideration of deregulation. For the first time since 1945, legislation provides subsidies for the provision of private rented housing, thereby enabling private landlords to compete on more nearly equal terms with other tenures, at a time when the supply of local authority rented housing is falling and its rents likely to be rising in real terms as HRAs are ring-fenced. The Local Government Act 1988 gives local authorities the power to provide financial assistance to private landlords letting assured tenancies to the extent of between 50 and 75 per cent of scheme cost, depending on location. More significantly, the Finance Act 1988 extended the provisions of the Business Expansion Scheme (BES) to rented housing. This provides significant inducements to private individuals to invest in private renting. To qualify for BES tax breaks up to 1993, individuals must invest a minimum of £500 and a maximum of £40,000 in any tax year, either directly or through a managed fund in unquoted property companies letting assured tenancies (assured shorthold letting does not qualify). This investment is eligible for income tax relief at investors' marginal tax rates and for full capital gains tax relief on disposal of shares after five years (although the company itself is liable for gains on property sales). Companies can invest up to £5 million per tax year (at least 80 per cent must be in property) and properties should cost no more than £85,000 outside London (£125,000 in London) at acquisition. There are no other targets, however, restricting the application of funds, neither in respect of qualifying properties nor tenants, unlike, for example, the low-income housing tax credit in the USA (Gruen, 1989). The legislation was introduced to give a 'kick start' to the revival of private renting. In other words, the tax incentives are justified as fostering a demonstration project.

Given that so much of the current private rented stock is sub-standard, it is also important to note that the government is restructuring the grant system for improving older housing. Legislation to give effect to its proposals has been incorporated in the Local Government and Housing Act. Standards are to be reduced. All grants to private landlords are to be discretionary. Grant paid will be related to a test of landlords' ability to finance the eligibility work out of rental and other income rather than the current system of fixed percentages of eligible costs.

Finally, it is also important to note the proposed ring-fencing of local authority housing revenue accounts and the revised terms of central government subsidy to such accounts. It is widely anticipated that these changes will lead to increases in local authority rents, and this may modify demand.

IMPACT OF DEREGULATION

The government has been right not to deregulate existing tenancies. Many tenants are elderly and need protection. Meanwhile landlords are getting a competitive return on the price they paid for the dwellings, since most will have been bought at sitting tenant prices. The tightening up of the laws on harassment and the possibility that illegally evicted tenants could get substantial damages should prevent the deregulation of new tenancies 'spilling over' to existing tenancies. In other words, the changes should reduce incentives to harass and illegally evict tenants in order to let on the deregulated sector or to sell with vacant possession. In any case the new legislation has not of itself increased incentives to do this, since landlords could already effectively let outside the Rent Act (if they got possession) before it.

One of the main problems for existing tenants is the need to get the physical conditions of their homes improved. In the long term, conditions will be improved when they leave (or die) and their properties are sold off to owner-occupiers. In the meantime it is important to them (and equally, if not more so, to the inner-city neighbourhoods of which they are part) that landlords have adequate incentives to improve and maintain their homes. The government's intention to place all grants to private landlords on a discretionary basis may reduce confidence amongst investors. Whether or not financial incentives are reduced depends on the specific rules introduced to estimate how much of the eligible cost of improvement should be funded from the rent income available to landlords. This will involve judgements about appropriate rates of return and about allowances for risk liquidity and management costs. None of these will be easy to estimate. Whilst the proposed system is likely to be more efficient in the use of public funds (since grants will no longer be available for work which can be supported by rents), it is less likely than the existing system to be equitable (since the discretionary basis could mean that grants could be awarded by local authorities in a capricious and arbitrary way unrelated to tenants' or landlords' needs), and it is possible that it will be less effective by increasing uncertainty and reducing incentives (Crook with Sharp, 1989). There is a case for a much more comprehensive approach than the one proposed, involving mandatory and discretionary grants together with tax allowances to encourage approved landlords to invest in run-down tenanted property and to bring it up to standard whilst protecting the interests of existing tenants.

The government's main concern, however, is about the ready-access sector, and its aim is to increase investment in this sort of accommodation. The overall intention of the legislative changes is to increase rents, to reduce the risk of this class of investment and to enhance its liquidity, thereby increasing returns to a competitive level and drawing in new responsible investors who will expand the overall supply and manage it prudently. Whether the government succeeds in its intentions depends, first, upon the extent of post-deregulation rent increases and the extent of demand at the rents landlords require. It depends, second, upon whether existing landlords decide to stay in the market and, more crucially, given the government's desire to attract responsible investment, whether new investors will be drawn in.

As far as demand is concerned, the government's targets are households at early stages in their life-cycle and job-movers. It does not intend private renting to provide for families and low-income households. The government is right here. Owner-occupation, local authority or housing association renting is better placed to meet the needs of families and the vulnerable. Moreover, given the existence of mortgage interest tax relief for owner-occupiers, renting at market rents without an equivalent fiscal or other subsidy for tenant or landlord is an attractive option for the consumer only under limited circumstances, such as people who move often and incur a lot of transaction costs when buying, high-income households not wanting the ties of home ownership, those who need to save up for a deposit to buy and those, like students, who are short-term residents (Coleman, 1989).

If there is to be an upward shift in rents it is unlikely that this will result from a shift in demand by existing tenants, many of whom will be too poor to pay higher rents. Indeed three factors suggest they will be facing greater difficulties than hitherto, particularly those just outside the Housing Benefit system.

First, students form a considerable proportion of new entrants. Their grants, however, were cut in real terms by 20 per cent between 1980 and 1987 and they have lost some of their eligibility to claim Housing Benefit; furthermore the government's announcement that grants are to be frozen and topped up with loans after 1990 increases uncertainty about levels of demand from students.

Second, the revised Housing Benefit scheme introduced in April 1988 worsened the position for those in work, those with occupational pensions and savings and those under age 25. Although recipients on Income Support have all their eligible rent and 80 per cent of their community charge paid (and whilst all recipients are protected from rent increases, all other things being equal), significant numbers, including low-paid workers, receive very little in help towards rents. Moreover, those on benefit are subject to high marginal rates of taxation because of the steep taper of 65 pence in every pound by which benefit is withdrawn when their net income exceeds their allowances (Ward and Zebedee, 1989).

Third, problems will also arise for Housing Benefit recipients when rent officers decide that the rent they are paying is not a reasonable market rent. If local authorities decide to exercise their discretion and pay benefit only up to the reasonable rent, tenants will then have to pay even more out of their own pocket or rents will fall, causing landlords to leave the market.

Thus, so far as the existing range of low-income tenants catered for by the private rented sector is concerned, there is little evidence that the market will sustain much higher rents, except in areas of pressure like London (although the formal lifting of regulation may, of course, remove the 'shadow' of the fair rent system on *de facto* deregulated rents and thus enable suppressed demand to express itself). Moreover, if rents do rise it will give further incentives for those in work and ineligible for Housing Benefit to transfer into owner-occupation where they will get tax relief on mortgage interest as of right. The legislation has done nothing to modify the attractiveness of owner-occupation, and increased rent amongst this latter group is likely to reduce demand for private renting.

There may be, however, new sources of demand willing to pay higher post-deregulation market rents. Two sources can be considered. First, upward pressure on local authority rents (combined with growing shortages of accommodation for applicants who are not homeless) may increase demand for private renting. It seems equally plausible, however, that any such rent increase will boost demand for owner-occupation rather than for renting alternatives. Second, it is plausible to expect an increase in demand from job-movers, well-paid professionals and other workers at early stages in their careers, provided there is an increase in the supply of new and well-improved houses to rent privately in desirable neighbourhoods. Such accommodation has been very difficult to find and many potential tenants who would otherwise value the flexibility that private renting brings choose to buy instead, not only because subsidised house ownership is very competitive with economic rents, but because housing of the quality and accommodation they want is simply unavailable. If it does become available they will retain flexibility, reduce transaction costs and not need to lock up capital in house purchase.

Whether or not there is such a supply increase depends, of course, on the extent to which the legislation has satisfied the conditions needed to keep existing landlords in the market and persuade new ones to come in (including its impact on lowering the cost of provision, for example, because finance is cheaper to raise within a deregulated framework).

It is generally accepted that four conditions are necessary to sustain and attract investment (for example, see HCEC, 1982). First, the rate of return must be competitive with alternative investments of similar risk and liquidity. Second, there must be greater and more predictable degrees of liquidity than in the past, so that landlords can be certain of the circumstances in which they can get vacant possession and realise the value of their assets. This is particularly crucial to landlords who are interested in getting a return from capital gain. Third, there must be stability in the new legal framework. If potential landlords expect a new government to reintroduce controls, they will not make investments. In other words, there must be confidence that deregulation will last. Fourth, the reputation of private renting as an investment needs to be sound. This means 'laying to rest the ghost' of Rachmanism. If landlordism is seen as a repugnant form of investment many people, particularly financial institutions, will not invest, even if the returns are good (Allen and McDowell, 1989).

These conditions do not apply equally to all types of landlord. It is generally thought that the larger, long-term investors – usually companies, including financial institutions – emphasise rent and competitive returns over the long run. By contrast, the smaller landlords – often individuals, not companies – are more concerned with the short run and with liquidity, and therefore security, and their ability to get vacant possession and secure capital gains (see HCEC, 1982; Price Waterhouse, 1989). In some respects it is possible to argue that assured shorthold tenancies were designed for smaller individual landlords and assured tenancies for larger, long-term ones. Although there is little hard evidence to date about the impact of deregulation, it is nevertheless possible to argue that whilst legislation has not satisfied all the conditions for investment by larger long-term landlords, it has gone some way towards doing so. Risk is significantly lowered because legislation now provides a *de jure* framework for letting at market rents with minimal security and reduces the risks of being saddled with 'bad' tenants. The rate of return required for this lower risk should be less than under the previous framework. Thus, given market rents, contractual arrangements for rent reviews in line with movements in prices and earnings, powers to remove bad payers and other rights enhancing liquidity, some of the requirements of long-term investors have been met by the legislation.

However, some important requirements remain to be satisfied. Whilst risk has been significantly modified, there has been no political consensus about the new framework and there is, as a result, no certainty about long-term stability. The existence of this political risk may be crucial in dissuading responsible institutions to make long-term investments. Indeed the legislation may need to survive a change in government intact before the risk that there will be changes to the law, especially the threat of the reintroduction of rent controls, significantly diminishes. Moreover, it is by no means certain that the legislation has increased the reputation of private rented investments. Reputation is particularly crucial to financial institutions and may have been undermined by the government's abandonment of the system of approving assured tenancy landlords introduced by the Housing Act 1980. Now any individual or company can let at market rents and contractual security. The government took the view that a system of approval would have high administrative costs and would not guarantee catching all 'cowboys'. Instead it preferred to police standards and significantly strengthened local authorities' powers to take the initiative in enforcing them.

On the other hand, it seems likely that more of the needs of smaller landlords have been met. Their time horizons are shorter than larger landlords with institutional funds and their interest in letting lies as much in capital gains as in rental returns. The legislation allows them to let at market rents, to remove bad payers and tenants easily, and to regain vacant possession when they want by letting on assured shorthold tenancies. It is probable, therefore, that most new private letting under the Housing Act 1988 will be of assured shortholds by small landlords. Given the significant reduction in their risk and their willingness to accept equity-related returns, such landlords could make competitive returns on current standards with no significant increase in rents.

CONCLUSIONS

It is evident from this analysis that, although the 1988 Housing Act has made significant progress in dealing with the problems of private renting, substantial problems still have to be tackled, especially about confidence and sustainable rents yielding competitive returns.

In relation to the latter it is also evident that additional subsidy is required if private renting is to be revived and social objectives achieved. The need for capital subsidy has already been demonstrated by the failure of the earlier assured tenancy scheme. This collapsed when capital allowances were withdrawn because investors considered returns were inadequate from rents at the then current levels of demand (Kemp, 1987b). The introduction of tax incentives under the BES is thus an important step forward, not only in giving investors competitive returns on their tax-deductible investments but also in improved reputation given the way some BES companies have arranged for housing associations and professionally qualified agents to manage their property. It remains to be seen, however, if the BES scheme will succeed in attracting long-term investment or only a short-lived wave of short-term speculative investment looking for tax-sheltered capital gains. The first signs from prospectuses of BES assured tenancy companies is that the investment will be short term, most emphasising capital growth (rather than rental returns) and 'exit routes' via property sales to enable investors to get tax-free capital gains.

Whilst some of the necessary conditions for private rentings revival have been put in place, important conditions remained to be fulfilled. It is likely that the 'central dilemma' of the Commons committee referred to above has not been fully resolved. Demand for private renting is likely to be mainly restricted to low-income households, and supply will be confined mainly to short-term, assured shorthold lettings by individual landlords in multi-occupied property at the bottom of the market. There may be some additional supply of small, newly built, reasonable-quality units triggered by BES incentives and a demand for these at competitive rents by job-movers and younger, better-off households not wanting to tie up capital in home ownership and attracted to private renting as a temporary staging post in their housing careers. But more generally many of the problems in private renting that have confronted tenants, landlords and local authorities in the 1980s are likely to endure into the 1990s unless significant steps are taken to increase demand by existing and potential tenants and create greater confidence amongst potential landlords.

If the competing tax and subsidy arrangements in other tenures are not to be reformed, some compensatory subsidy to private renting is required. The BES scheme is a start. Capital subsidies should be allied to a system of prior approval of landlords. This is likely to increase bipartisan support. Local authorities' discretionary and mandatory powers to police and enforce standards should be extended. Retention of approved status (and therefore tax breaks or other subsidies and the right to let at market rents) would depend on letting at acceptable minimum standards. Standard contracts for tenancies should be introduced and housing courts set up to provide speedy arbitration of disputes. The Housing Benefit scheme needs to be remodelled, allowances increased and tapers on benefit should be made less steep.

All of this needs more rather than less public investment in the support of private rented housing. Without it, sustainable investment in habitable housing on secure terms and at affordable rents is most unlikely to occur in the 1990s.

POSTSCRIPT

Since this chapter was written in 1989 a number of developments have occurred. The purpose of this postscript is to discuss these developments and more recent research undertaken on private rented housing since deregulation and their implications. First, I examine evidence of an increase in supply and demand, secondly, the impact of tax incentives and thirdly, the renewed interest in the revival of private renting and the need for subsidies.

Supply and demand

First, there is some evidence to suggest that the rate of decline of the sector may have slowed down since 1988 and that there has been a small growth in lettings in southern England, especially in London (London Research Centre, 1990; Office of Population Censuses and Surveys, 1991; Franklin, 1991).

However, it is not clear if this increase in lettings can be attributed to deregulation or to the recent slump in the property market in Britain – or to both. It is certainly possible to argue that recent changes in the housing market have increased both the supply of and demand for private rented housing.

On the supply side, the fall in demand for houses to buy has been associated with a nominal as well as a real fall in house prices, especially, though not only, in the South-east of England. This has left home owners who have been moving for job and other reasons unable to sell their houses. Rather than sell for prices significantly below their expectations (and/or their outstanding mortgages), some of these owners may have rented them out temporarily on shorthold, in the meantime moving house themselves, awaiting an upturn in the market before making renewed attempts to sell. At the same time, building societies have been auctioning large numbers of properties repossessed from those unable to pay their mortgages and some of these may have been bought by landlords at prices which make renting attractive, especially in the light of the capital gains realisable given the anticipated eventual upturn in the market. Housebuilders with unsold stock may have similarly rented out property. The scale of activity, however, is unknown, since the evidence is partial and of a case-study nature. As a result, it is not possible to say whether this is on a significant scale and more than a temporary phenomenon.

It is also possible to argue that recent conditions have been equally conducive to an increase in demand for private renting, particularly amongst those who have positive reasons for doing so. These would include households starting their housing careers, young singles, especially students, job movers, moving owner-occupiers wanting to rent temporarily, and households splitting up through divorce and separation. None of these groups will want to rent accommodation in any one place for a long time but all have positive reasons for valuing the flexibility that private renting can give them. It might be expected that recent housing-market and other changes have reduced the demand for owner-occupation amongst these groups. Firstly, the recent uncertainty about house prices increases the risks for mobile households. Secondly, the relative investment attractiveness of owner-occupation, compared with other investments, has fallen (e.g. TESSAs, modifications of mortgage interest tax relief, indexation of capital gains). Thirdly, deregulation has created a wider range of contracts than were previously available. Fourthly, deregulation has produced, via the Business Expansion Scheme (see below), good-quality, private rented housing in desirable areas that has not existed for many years.

All this suggests that recent conditions appear to have favoured an increase both in the supply of and demand for private renting, particularly amongst upper income groups, but it is far from clear whether these changes have been on a large scale and more than a temporary response to the property slump.

Meanwhile these is some evidence that conditions at the bottom end of the market have not improved, as anticipated in the chapter above, especially in areas of housing pressure like Greater London (Sharp, 1991).

Business Expansion Scheme

As the chapter above explained, the provisions of the Business Expansion Scheme (BES) were extended to companies raising money to let assured tenancies, not as a way of permanently providing a subsidy to private landlords but of mounting a demonstration project of the renewed profitability of private renting that was created by deregulation. It was the government's belief that the market would produce competitive returns. It accepted, however, that it would take time for confidence to return and that for the short run there would be some risk in investing in private renting because levels of demand and rents were uncertain, so too were management and maintenance costs, voids, bad debts, capital gains and therefore yields. There would still be a perception of risk, especially amongst the potential new investors that were the target of policy. It would be difficult therefore to get private finance in, except at a very high risk premium, and the BES has been used to provide generous tax incentives to attract private finance for a short-term demonstration project.

Research on the impact of the BES on the provision of private rented housing shows that it has produced a small revival in the sector but that this is likely to be short-lived (Crook *et al.*, 1991). New companies have been set up to raise funds for the BES and they have generated the first significant wave of new investment in private renting for half a century. The dwellings acquired by BES landlords are of a much higher quality than those in the rest of the private rented sector. This increase in supply was almost wholly additional. Company directors doubted if they could have raised debt or equity capital without the benefit of subsidy.

Despite this impressive achievement – achieved at a substantial cost in tax expenditure – it is doubtful if the success will be long-run. Although many of the companies' directors would like them to continue letting residential property in the long term, few of them thought that this was likely to happen. The available evidence suggests that their net returns are not competitive (Crook and Kemp, 1992). Rents, net of costs and capital gains, are 5 per cent of vacant possession values and these returns do not appear sufficient to attract private finance on a large scale without subsidy. It is likely therefore that after five years, when the initial shareholders will want to realise their tax-free capital gains, each company will decide to sell off its properties with vacant possession into the owner-occupied market.

Future prospects

The objective of deregulation is the creation of a healthy and profitable private rented sector which provides habitable, reasonably secure and affordable housing to tenants with positive reasons for being private tenants. An additional objective is to draw in large-scale private financing and in particular to encourage new companies to enter the market and to create a more modern form of landlordism or 'new model landlords'.

The main lesson of deregulation so far is that whilst much has been done to get the framework right for achieving this, particularly with respect to raising confidence, the main difficulty lies on the demand side. Whilst there might be reasons to expect a transfer of demand from upper-income groups currently in the owner-occupied sector (for reasons outlined above), the main burden of the evidence, including the BES research, is that market rents alone will not produce competitive returns and pull in large-scale private finance.

All this confirms the need for subsidy for private landlords, whilst current arrangements which provide subsidies for social landlords and owner-occupiers endure. It is in this context that the BES experiment has made two very important contributions to the revival of private renting. Firstly, it has reintroduced the venture capital industry to private renting and secondly, and most importantly, it has made it possible to discuss subsidies for private landlords as politically viable options.

The regeneration of private renting is now firmly on the policy agenda. It is increasingly accepted that the country needs a viable private rented sector. The recent difficulties in the owner-occupied housing market during the property slump, together with the constraints on increasing the social rented sector, will have reinforced the acceptance of this need amongst representatives from all political parties. As a result, there is now widespread acceptance of the need to find an appropriate way to revive private renting.

Whilst there is by no means universal agreement on how this should be done, a number of recent reports have suggested ways of injecting subsidy into the sector, some of the proposals being part of wider packages of housing finance reform (Crook *et al.*, 1991; Hills, 1991; Inquiry into British Housing, 1991; Maclennan *et al.*, 1991; Merrett, 1991; see also Crook and Kemp, 1992, for a review and comparison of these proposals).

These proposals and the debates surrounding them and the future of private renting suggest that there is now a much greater consensus on the need to promote private renting and on some of the mechanisms to achieve this (see Best *et al.*, 1992). But despite the evidence about returns being achieved in the deregulated market and despite the arguments presented in the reports referred to above, the government remains committed to relying on deregulation without the provision of subsidy.

The case for resolving the Commons' committee's central dilemma is still clear ten years after their report was published and, moreover, is well supported by the evidence of the period since deregulation. The political climate for resolving this dilemma by the provision of subsidies is now much more favourable. Whether the public-expenditure climate and the government's ideological commitment to the market will allow them to resolve this dilemma is much less clear.

REFERENCES

Allen, J. and McDowell, L. (1989) *Landlords and Property, Social Relations in the Private Sector*, Cambridge: Cambridge University Press.

Arden, A. (1989) *Manual of Housing Law*, London: Sweet & Maxwell.

Bentham, G. (1986) 'Socio-tenurial Polarisation in the UK 1953–1963: The Income Evidence', *Urban Studies*, vol. 23, pp. 157–62.

Best, R., Kemp, P., Coleman, D., Merrett, S. and Crook, T. (1992) *The Future of Private Renting: Consensus and Action*, York: Joseph Rowntree Foundation.

Bovaird, A., Harloe, M. and Whitehead, C. (1985) 'Private Rented Housing: Its Current Role', *Journal of Social Policy*, vol. 14, no. 1, pp. 1–23.

Coleman, D.A. (1989) 'The New Housing Policy – A Critique', *Housing Studies*, vol. 4, no. 1, pp. 44–57.

Committee on Housing in Greater London (1965) *Report*, Cmnd 4609, London: HMSO.

Crook, A.D.H. (1986) 'Privatisation of Housing and the Impact of the Conservative Government's Initiatives on Low Cost Home Ownership and Private Renting between 1979 and 1984 in England and Wales', 4, Private renting, *Environment and Planning* A, vol. 18, no. 8, pp. 1029–37.

——(1989a) 'Investment in Private Rented Housing: Evidence from Sheffield', in Cross, D. and Whitehead, C. (eds) *Development and Planning*, Cambridge: Policy Journals.

——(1989b) 'Multioccupied Housing Standards: the Application of Discretionary Powers by Local Authorities', *Policy and Politics*, vol. 17, no. 1, pp. 41–58.

Crook, A.D.H. and Martin, G.J. (1988) 'Property Speculation, Local Authority Policy and the Decline of Private Rented Housing in the 1980s', in Kemp, P. (ed.) *The Private Provision of Rented Housing*, Aldershot: Avebury Books.

Crook, A.D.H. and Kemp, P.A. (forthcoming) 'Reviving the private rented sector?', in Gibb, K. (ed.) *United Kingdom Housing Finance and Subsidy*, Aldershot: Avebury Books.

Crook, A.D.H. with Sharp, C.B. (1989) *Property Dealers, the Private Rented Sector and Local Authority Policy in Urban Areas of the North and Midlands*, occasional paper, Sheffield: University of Sheffield, Department of Town and Regional Planning.

Crook, A.D.H., Kemp, P.A., Anderson, I. and Bowman, S. (1991) *Tax Incentives and the Revival of Private Renting*, York: Cloister Press.

Department of the Environment (1987) *Housing: The Government's Proposals*, Cmnd 214, London: HMSO.

——(1988) *English House Condition Survey 1986*, London: HMSO.

Doling, J. and Davies, M. (1984) *The Public Control of Private Rented Housing*, Aldershot: Gower.

Franklin, A. (1991) 'The private rented sector in Bristol', Working Paper 100, Bristol: School for Advanced Urban Studies.

Gruen, N. (1989) *The Low Income Housing Tax Credit and Private Sector Investment in New Rehabilitated Affordable Housing*, San Francisco: Gruen Associates.

Hamnett, C. and Randolph, B. (1988) *Flat Break Up and the Decline of Private Renting*, London: Hutchinson.

Harloe, M. (1985) *Private Rented Housing in the United States and Europe*, Beckenham: Croom Helm.

Hills, J. (1991) *Thirty-nine Steps to Housing Finance Reform*, York: Joseph Rowntree Foundation.

Holmans, A.E. (1987) *Housing Policy in Britain*, London: Croom Helm.

House of Commons Environment Committee (HCEC) (1982) *First Report Session 1981–82. The private rented housing sector*, HC4Ol, London: HMSO.

Inquiry into British Housing (1991) *Second Report*, York: Joseph Rowntree Foundation.

Kemp, P. (1987a) 'The Reform of Housing Benefit', *Social Policy and Administration*, vol. 21, no. 2, pp. 171–86.
——(1987b) 'Assured Tenancies in Rental Housing', *Housing Review*, vol. 36, no. 2, pp. 43–5.
——(1988) 'New Proposals for Private Renting: Creating a Commercially Viable Climate for Investment in Rented Housing?', in Kemp, P. (ed.) *The Private Provision of Rented Housing*, Aldershot: Avebury Books.
London Research Centre (1990) *Renting Revival. A Study of the Impact of the 1988 Housing Act on Islington's Private Rented Sector*, London: London Research Centre.
Maclennan, D. (1988) 'Private Rented Housing: Britain Viewed from Abroad', in Kemp, P. (ed.) *The Private Provision of Rented Housing*, Aldershot: Avebury Books.
Maclennan, D., Gibb, K. and More, A. (1991) *Fairer Subsidies, Faster Growth*, York: Joseph Rowntree Foundation.
Mallinson, M. (1989) 'The Producers', paper presented to Conference on Affordable Housing, London 1989, sponsored by Joseph Rowntree Memorial Trust (mimeo).
Malpass, P. and Murie, A. (1987) *Housing Policy and Practice*, 2nd edition, London: Macmillan.
Merrett, S. (1991) *Quality and Choice in Housing*, London: Institute for Public Policy Research.
National Federation of Housing Associations (NFHA) (1985) *Inquiry into British Housing*, London: NFHA.
Nevitt, A.A. (1966) *Housing Taxation and Subsidies*, London: Nelson.
Office of Population Censuses and Surveys (OPCS) (1987) *General Household Survey, 1985*, London: HMSO.
——(1991) *The 1990 Private Renters Survey: Preliminary Results*, OPCS Monitor SS 91/2, London: HMSO.
Paley, B. (1978) *Attitudes to Letting*, London: HMSO.
Price Waterhouse (1989) *Study of the Private Rented Housing Market*, London: Price Waterhouse.
Sharp, C. (1991) *Problems Assured*, London: Shelter Housing Aid Centre.
Thomas, A.D. with Hedges, A.(1986) *The 1985 Physical and Social Survey of Houses in Multiple Occupation in England and Wales*, London: HMSO.
Todd, J., Bone, M. and Noble, I. (1982) *The Privately Rented Sector in 1978*, London: HMSO.
Todd, J. and Foxon, J. (1987) *Recent Private Lettings 1982–84*, Technical Appendix, London: HMSO.
Ward, M. and Zebedee, J. (1989) *Guide to Housing Benefit*, London: Institute of Housing and SHAC.
Whitehead, C.M.E. and Kleinman, M.P. (1989) *Private Rented Housing in the 1980s and 1990s*, Cambridge: Granta Editions.

6 The 1987 housing policy
An enduring reform?

D.A. Coleman

INTRODUCTION

Before the 1987 general election the Conservative government identified rented housing as the next area for reform, following the success of its earlier policies to encourage private ownership and to give council tenants the right to buy their homes.

Low standards, inadequate choice and dissatisfaction with housing in Britain were and remain concentrated in the rented sector. The £20 billion backlog of repairs in the public sector (DoE, 1985) and the concentration of disrepair, overcrowding and defective amenities in the private sector were well known (House of Commons, 1982; HRH the Duke of Edinburgh, 1985; DoE, 1988d). Local authority housing has particular problems of its own (Coleman, 1985; Power, 1988). Its management was evidently beyond the capacity of many local authorities (Audit Commission, 1986). Council rents have been sometimes kept artificially low, or not efficiently collected, as a means of patronising the local electorate. The experimental designs favoured in some post-war council housing estates were frequently technically defective and socially harmful. They tend to isolate and stigmatise their tenants. The government believed then and now that the social and economic problems of inner cities are exacerbated by some urban estates and the attitudes it believes they encourage. Council housing lies at the end of a queue which some, particularly the young and single, may be unable even to join.

Private renting has been marginalised by decades of rent control and security of tenure. Its remnants usually offer only a restricted range of accommodation, at near black-market rents in areas of high demand, with transient households preferred. It is now seldom the first choice for anyone; it is regarded as an unsatisfactory intermediate before home ownership. From a financial point of view it has not been a rational long-term choice for families for decades. By the mid-1980s, the controls on rents had begun to look particularly anomalous after several years of deregulation and privatisation since 1979, and no Conservative Party Conference passed without calls for their removal. In all these respects Britain has been unlike most other industrial countries, where housing tenure is more a matter of choice or convenience, not a foregone conclusion. Distortions of cost and supply were believed by the Conservatives to contribute substantially to homelessness, unemployment and local labour shortages by inhibiting labour migration, and to wasted resources by keeping too many properties empty.

At the general election of 1987 all three major parties put forward proposals to improve standards and choice in public renting. The Conservatives and the Alliance wanted to expand private renting too, in different ways, while the Labour Party maintained its opposition to any revival of the private rented sector (with the possible exception of old-style assured tenancies), maintaining its long-standing view that abolition of private renting was the most desirable course. It was the Conservatives who had the opportunity to put their policies into effect.

The arguments for and against the government's proposals have been extensively debated in the political and academic arenas. The purpose of this paper is not to take that debate further, but to consider the reasons which persuaded the government to make such radical proposals in such a potentially politically dangerous area and to see how far the Housing Acts and related measures seem likely to realise their aims, within their own terms of reference. It is not a review of housing policy or its development (see instead Burnett, 1986; Daunton, 1987; Holmans, 1987).

The complex question of mortgage income tax relief is not considered here, partly because there is no immediate prospect of any change in policy. Government policy of unqualified support at the basic rate of tax remains unchanged although supporters of the relief accuse the government of allowing it to wither by keeping it fixed at £30,000, well below the average house price of £53,000. In March 1988 the Chancellor of the Exchequer announced that from August 1988, multiple relief would no longer be available for the joint purchase of the same property by unmarried individuals, a privilege which had the anomalous effect, for a 'party of the family', of giving more support to communes and cohabitees than to married couples. Other critics blame tax relief for underwriting the inflationary growth of house prices and inflation in general (see Bover *et al.*, 1988) and distorting savings patterns, a view increasingly widely shared in all parties and held by commentators as varied as the International Monetary Fund (1989) and Norman Tebbitt (article in *The Times*, 26 September 1989). In 1987, opposition parties did not go further than to promise the removal of relief from higher rate tax, which was in any case implemented in the 1991 budget. The Liberal Democrats alone promised the replacement of mortgage income tax in 1992. None has dared to suggest the reform of another tax incentive to investment in housing, namely the exemption from capital gains tax on the sale of the principal private residence. These tax reliefs are discussed tangentially in connection with the problems of the revival of renting. The additional pressure on house prices and rents generated by the restrictions on the supply of building land created by the planning system, particularly in areas of high demand in the southern part of the country (see Evans, 1988), is discussed in a later section. There seems to be growing evidence that the distortions in price caused by these reliefs and controls are acting neither in the interests of house-buyers nor the economy in general.

This chapter will concentrate on the rented sector, to which most recent legislative changes are directed, and attempt to evaluate the policies' likely success. It concentrates on the legislative changes introduced in 1988 and 1989, and does not comment on the minor additional proposals on renting put forward by the Conservatives at the 1992 General Election. These do not look as if they will radically affect the situation one way or another. Success will be measured by the answers to the following questions:

1 Will the decline of private renting be halted?
2 Will private money go into new building for private rent?
3 Will major institutions, especially the building societies, become heavily involved in new private renting and in partnership with housing associations?
4 Will enough private money go to housing associations to expand a genuine alternative 'social renting' sector?
5 Will tenants agree to transfers out of local authority control, or choose a new landlord?
6 If they do, will the change mean more than a continuation of council tenancy by other means?

PROBLEMS AND POLICY RESPONSES

When the Conservatives were returned to power in 1987 they faced several obstacles to a successful reform of renting:

1 Political opposition to any interference with the vested interests of existing tenants, and widespread public suspicion of the existing private sector in rented housing;
2 Lack of interest by most potential investors (except building societies) in providing housing for rent;

3 A fiscal environment which made it difficult for private renting to compete with home ownership;
4 Planning controls which inflated prices of new housing, rented or otherwise, in areas of greatest demand;
5 The lack of a clear, rational system of support for low-income households (low rents for all council and housing association tenants and individual housing benefit for the majority of tenants both contributing to subsidy in a confusing way); and
6 The absence of a plausible private-sector alternative to housing associations for the proposals to devolve local authority housing.

The following section will consider how far these obstacles are likely to be removed by current policies.

HOUSING AT MARKET RENTS

Renting is not one market; it is several. These markets should not be regarded as being separate from ownership but as overlapping it to some degree. But to begin it may be useful to regard it as having two major divisions: one which overlaps ownership for those on average or above-average incomes, most of whom will eventually buy; and another, sometimes called 'social renting', for those whose incomes are inadequate for ownership or for an economic rent and which in many cases will remain so for all their lives.

Let us look at the 'economic' market first. Several obstacles may impede rapid reversal of the decline in the supply of private renting. Many landlords, especially the more law-abiding ones, have waited a long time to dispose of rented property made uneconomic and troublesome by rent controls and by the effectively permanent tenure enjoyed by many tenants. It will take clear examples of success to restore their confidence. By themselves, market rents on new tenancies may encourage only a modest increase in new rental housing because the market will remain dominated by the relative cheapness of buying with a mortgage compared with renting (there were 9.6 million outstanding mortgages in 1991–2). Mortgage income tax relief is available on the interest payable on a mortgage up to a maximum value of £30,000 per dwelling. For the average borrower in 1991–2 this subsidy was worth about £640 per year and £900 for first-time buyers.

In 1992 the average mortgage outstanding was £27,000; in the last quarter of 1991 the average new building society mortgage was £39,400 for first-time buyers and for others £49,500. The total cost of tax relief varies with interest rates: £7.7 billion in 1990–1, £6.1 billion in 1991–2. The average relief per mortgage fell from £820 to £640 (Social Trends 20 and 22, section 8; Boléat, 1989; *Housing Finance* 14, May 1992). In general, relief tends to cushion the effects of higher interest rates because the tax rises with them. This tends to weaken attempts to control house price inflation by rising interest rates. This relief generally benefits better-off households more in absolute terms than poorer ones, and non-taxpayers not at all. The option mortgage scheme, now absorbed in the Mortgage Income Tax Relief at Source (MIRAS), benefits lower-income taxpayers who only pay tax on a small part of their income. In addition, house purchase is made a very attractive investment by being exempt from capital gains tax (CGT) on the sale of the principal residence. Landlords are not so exempt and cannot 'rollover' CGT on the purchase of new property, but they can obtain tax relief on loans for the purchase of property to let.

However, renting at market rents, even without an equivalent fiscal or other subsidy, may be an attractive or cheaper choice for the consumer under limited circumstances. For example:

1 It is easier to rent than to buy a smaller volume of housing and thereby live more cheaply in central or fashionable urban areas (bed-sits, flat-share).
2 Repeated moves by mobile people may incur such transaction costs through changes of ownership that market renting is cheaper;
3 For some people (usually on high income), avoiding the chores of home maintenance and repair may be worth the cost of renting.
4 Would-be purchasers without a 100 per cent mortgage must wait to save a deposit.
5 Only renting makes sense for short-term residents, including students and businesspeople on short-term secondment from other regions in the UK or overseas.
6 In addition to these components of demand, market rents and the new assured shorthold tenancy with guaranteed repossession, even without subsidy, should encourage landlords with vacant property which they do not want to sell immediately to let once more. Premises above shops and offices and houses difficult to sell in the present house price slump are obvious examples; Business Expansion Scheme (BES) companies have been a particularly attractive outlet for the latter.

A few renters might choose to rent at market levels for most of their lives, perhaps because some or other of the circumstances above have become permanent. But in Britain, for most people, renting has become a transient category between leaving home and buying a house. In 1989, the latter was the ambition of the overwhelming majority (over 81 per cent) of households (see Boleat, 1989, ch 6). Private renting, to single people or couples who expect eventually to own their homes, is perhaps the category capable of the biggest expansion if the supply of more acceptable dwellings to rent is increased by the 1988 Act. It might enable home ownership to be deferred. A high proportion of households eventually becoming home-owners is quite compatible with a longer period spent renting than is the case at present (as in the USA). UK home-ownership rates for people in their twenties are the highest in the Western world. This is partly because private renting has been so comparatively unattractive and inaccessible. The Property Owners Federation have been pessimistic about prospects, and traditional landlords do not seem prominent in the recent BES revival.

So far there is no systematic information on changes in the supply and demand for rented housing following the removal of controls from new lettings. However, newspaper reports suggest that some owner-occupiers are encouraged by the new rent regime and spurred by the present difficulties of selling property to offer their houses to let. Estate agents' reports on changes in demand and supply are mixed (*Sunday Times*, 24 September 1989; Best *et al.*, 1992), but some report strong demand. One obvious new change is the proliferation of prominent advertisements for rented homes by estate agents and others on, for example, the London Underground, at a level not seen since the 1930s.

The intensity of demand in London even for existing unsatisfactory private renting shows there is some room for expansion of an 'orthodox' free-market sector without further compensation for landlords. This is because major cities, and particularly London, continue to attract young immigrants in the 16–24 age-group, despite their general tendency over the past few decades to lose population. Furthermore, London in particular has a booming service economy and house prices are out of reach of people at the bottom of their earnings curve. According to some reports, about a third of households in the southeast who are not yet home-owners have incomes inadequate for the purchase of mortgageable property (ADC, 1989). London has the highest proportion of households living in private rented accommodation: about one in six (Whitehead and Kleinman, 1987).

Even before the new legislation, new tenancies, in London and elsewhere, were usually already offered at 'market' rents through various techniques for avoiding the Rent Acts: as company or holiday lets, or through licence, or informally without reference to any legislation. A majority were furnished lettings (which command a higher rent than unfurnished) and a high proportion were let by resident landlords (Todd *et al.*, 1982; Todd, 1986; GLC, 1986). Supply was nonetheless restricted. Under the previous law (which continues to apply to existing tenancies) the status of a tenancy is seldom certain without a court determination. Landlords run the risk of having their tenancies declared 'protected', the rents thereby reduced and the tenure made more or less perpetual. So for most landlords the ideal tenant has been one who is unlikely to want to stay for long. Outside the luxury market, or genuine company lets, little accommodation has been offered which is suitable for couples or families. It has been suggested that rents in London and other areas of high demand are at above open-market levels because of the restrictions on supply; and that the black-market rents charged include an element of premium to compensate the landlord for the risk of falling into rent control and for the difficulty of operating in the limited legal space in which private tenancies can be offered. An uncomplicated system of market rents should help this sector to expand. But the scale might be of the order of a few tens of thousands of tenancies, not the million or more by which it might eventually expand if private renting were also made competitive by the tax system or some other countervailing system of subsidy. This has been the almost unanimous view of potential landlords and investors (e.g., Halifax Building Society, 1987; Melville-Ross, 1987; Leeds Residential Property Association, 1987). The comparison between the 1988 and 1990 Private Renters Surveys suggests that while old regulated tenancies have declined fast, enough new assured and assured shortheld tenancies have been created almost to arrest the overall long-term decline in private renting (OPCS, 1991; Coleman, 1992).

Subsidies for rented housing

While mortgage income and capital gains tax relief remain in place, and while low-rent housing is offered to poorer households, two tiers of subsidy are needed to put private renting on an equivalent financial footing. The first, for the 'economic' market, is to compensate for mortgage income tax relief. The second – which might be a quite different mechanism, possibly grant aid rather than tax relief – is needed in addition to allow the private sector the same privileges as the public sector in offering tenancies at below economic rent. Successful private rental systems in other countries, such as those in the US and former West Germany, usually operate on a rough balance of fiscal or grant advantage with owner-occupation (see Stahl and Struyk, 1985; Hallett, 1988). Encouragement in investment – especially new building – for renting is encouraged by a variety of tax shelters, guaranteed or low-interest loans and generous depreciation allowances, especially for building for rent to low-income families.

Even after the 1986 US tax reforms there remain substantial advantages in building for rent in that country (National Association of Home Builders, 1986). In former West Germany the subsidy for owner-occupation is much more limited and is well balanced by tax advantages for letting property, especially for small landlords. In many of these societies (but not all – e.g., the Netherlands) looser planning controls make new development easier in areas of high rent or price – although often at an environmental cost.

The obvious approach to fiscal equity would be to moderate or phase out mortgage income tax relief. But as this is unlikely to happen in the forseeable future, alternatives must be considered. The government recognised the disability suffered by market renting and responded in 1988 by providing a counterbalancing subsidy for renting – the Business Expansion Scheme (BES).

The Business Expansion Scheme and the private rented sector

The main tax advantages under the BES, which was originally set up in 1983, are considerable: they include tax relief at the highest rate paid by the investor (i.e., up to 40 per cent) with a maximum investment of £40,000 per year gross (i.e., £24,000 net) and a minimum of £500; no capital gains tax when the shares are sold (although CGT is payable on the sale of the property held by the company). In addition, so called 'closed' companies of not more than five investors could retain tax relief on borrowing for further investment until that privilege was stopped in 1989. There are, however, a number of restrictions on BES investments. For example, the company must not be quoted on the Stock Exchange or the Unlisted Securities Market. Investors qualifying for the relief must not own more than 30 per cent of the company, or be employees or directors of it. Full tax relief requires the shares to be held for five years, and no individual or company may control a BES company (control may include loans as well as shares).

The Finance Act 1988 extended the scheme to companies involved in private renting. To qualify, a company must specialise in the letting of residential property on assured tenancies as a business enterprise for at least four years from the issue of the qualifying shares. The tenancies offered must be full assured tenancies (i.e., with security of tenure), not assured shorthold. The property must have all the standard amenities and not be statutorily unfit. The market value of each dwelling let (each house or each flat in a house) must not exceed £125,000 in Greater London or £85,000 elsewhere. The tenancies must be new, but the dwellings can be new or already existing. The relief is not transferable to another company if the property is sold.

Until the 1988 Finance Act there was no limit on the amount of BES equity that a company could raise in one year. But that Act imposed a limit of £500,000 on each BES scheme except rented housing and shop-leasing where the limit is £5 million per scheme. This may further encourage renting schemes by diverting funds which might otherwise have gone into conventional BES schemes. Further details are given in the 1988 Finance Act, in Inland Revenue publications (1987, 1988), and in Coleman and Bloch (1988).

The BES scheme has attracted much interest in investment in renting, for the first time since the Second World War. The terms are generous – giving perhaps an effective 17 per cent rate of return, depending upon what assumptions are made about rent levels and the discount on properties let at market rents, which will determine the value of the assets. The most popular estimate of this discount seems to be about 15 per cent. In 1988 BES promoters Johnson Fry speculated that the investment in the 1988–89 tax year would be about £500 million, rising to £1.5 billion the year after, compared with a total of £180 million for all BES schemes together in 1987–88.

Up to April 1990, £461 million was actually invested in BES assured tenancy schemes (Crook *et al.*, 1991). The average value of the dwellings involved is about £53,000. This yields about 9,700 rented dwellings. This looks like small beer compared with the perhaps 1.3 million dwellings still within the private sector and the total of about 200,000 built each year. Its importance lies in the fact that this is the first substantial new building or acquisition for rent since the war and through its attendant publicity may, if it succeeds, help re-establish the renting option as a more normal option before, or instead of, ownership.

Relatively few schemes operate in London despite the strong demand. The low capital value limit is a deterrent. Most schemes are in provincial towns, especially the north and Scotland where rental yields are estimated at 12 per cent because of the low cost of property. Prospects for the 1989–90 tax year were more modest because the tax relief could no longer reach back to the 60 per cent level and the borrowing relief which 'closed companies' could enjoy was abolished. The effects of the house price slump may have increased interest in BES because of the importance of capital growth in estimates of the final yield.

Some schemes have co-operated with housing associations to deal with the management of the property. The latter are the only substantial source of housing management expertise outside the local authority sector. Some associations were suspicious of this private-sector-oriented innovation, although their National Federation explored ways of using BES to help house their traditional target group of low-income families, and have given the scheme a cautious welcome. A Newcastle-based association, Nomad Housing Group, was the first to provide development and management services to a BES company. The limit of £5 million per scheme and the personal nature of the BES relief seemed to preclude direct investment by the most likely investor of all, the building societies. But in September 1988 and 1989 the Nationwide Anglia Building Society launched major BES initiatives on rented housing, in which it has long proclaimed an interest, by brigading together a large number of separate schemes.

BES is not an ideal choice for encouraging new rented housing. The choice of BES, out of all the other options, probably stems from the present government's aversion to creating new tax exemptions or other asymmetries in the tax structure, which it insists distort the price mechanisms and interfere with market forces. Mortgage interest tax relief has stood firm against this principle and no doubt will continue to do so. But there is a strong disinclination to introduce new exceptions. The logical neccessity of some kind of counterbalancing subsidy has required the government to take some action to encourage its policy on renting, and the BES scheme had the great advantage of being there already as part of a battery of schemes to encourage the creation and growth of small businesses.

One of the drawbacks of BES schemes is that profits can only be taken by withdrawing from the scheme after the minimum period of five years. Therefore BES allows money to be made from getting into renting only by getting out of renting again after a few years. But to rebuild public confidence in renting, a long-term commitment is needed. It will be difficult to revive it unless mechanisms are developed which permit commercially competitive rates of return from long-term renting and which develop the necessary management skills. BES does not do this. However, it does help to resolve part of the conundrum by getting things started, because long-term commitment, however necessary, is least likely at the beginning of rent reform when potential investors are most nervous that their investments might effectively be confiscated by the re-imposition of rent and other controls. In this respect the short-term nature of the incentive is an advantage in a politically risky area long starved of funds. But it was announced in the 1992 Budget that BES was to end in 1993 and no replacement scheme to encourage private rented housing has yet been announced.

It is still not clear, and cannot be for some years, how the shares will be disposed of and to whom. Selling the houses should be no particular problem. There are already a number of private landlords actively buying tenanted property at a discount. By the time that the first wave of BES expires, new renting ventures such as Quality Street and its followers may be in a position to take over property. Amalgamations of existing schemes are likely. Some housing associations may be interested in acquiring the properties even though the tenants are unlikely to belong to their usual client groups. The value of the shares will depend on the value of the properties and also on the nature of the successor institutions to the BES companies. There has been some speculation that BES companies may be tempted to put pressure on tenants to leave so that houses can be sold for owner-occupation, without a discount. New tougher measures, including stiff compensation in the 1988 Housing Act, should prevent overt harassment. But it would be legal to insert high rent increases into the assured tenancy contract to come into effect near the end of the BES period. If this were practised on a large scale and threatened to drive a lot of the new dwellings out of renting, there could be strong political pressures upon the government to take action to stop its policy being discredited. The political risk was considered to be substantial. The Labour Party was still committed to rent control in its 1989 statement on housing policy (Labour Party, 1989), although it had considerably modified its stance, accepting the need for some private renting (but not BES) in its manifesto in 1992.

Other grants and subsidies for landlords

Existing subsidies could not replace BES as an eventual long-term subsidy for renting, although they might usefully supplement it. Housing Benefit is a consumer subsidy paid to all kinds of tenants on the basis of their income, household and rent level (totals £2.3 billion in 1979–80, £4.6 billion in 1990–1). It is a complex system and was reformed in 1988 (for the pre-1988 system see Lakhani and Reed, 1987; for the new system see Ward and Zebedee, 1988). About 45 per cent of private tenants, 60 per cent of council tenants and 70 per cent of housing association tenants receive Housing Benefit to pay for all or part of their rent – in all about 5.18 million households in 1988 (6.85 million in 1987). In the public sector alone, this cost £2.73 billion in 1988–89; together with £1.2 billion in Exchequer and Rate Fund contributions, this amounted to £680 per public-sector dwelling. Although it is not paid to landlords it helps some landlords indirectly by increasing effective demand and the level of affordable rents. Some small landlords specialise in letting to people dependent on Housing Benefit.

A more direct producer subsidy has been made possible through Section 28 of the Local Government Act 1988. This gives local government new discretionary powers to give capital grant to investors in new or refurbished private housing for rent. The risk and the ownership must remain in the private sector. It is part of the policy of turning councils from housing imperialists into 'enablers'. Initially the grant level was 30 per cent, like the pilot scheme for Housing Association Grant (HAG) to attract private finance into housing associations. But after complaints that such a level of subsidy was insufficient for viable projects at affordable rents, the Section 28 grant level has been set at the same level as the 'variable' HAG, namely 50–75 per cent of capital cost. But it is up to the local authorities to decide what to do, and the expenditure must come from their own resources. Some local authorities may welcome the idea of discharging some of their housing responsibilities by subsidising the private sector and getting more housing built per unit of expenditure. Others, such as Labour authorities in inner-city areas, are likely to have strong ideological objections to subsidising any kind of private landlord. By 1992, this grant had been very little used. There are also avenues for subsidising appropriate housing projects through City Grant and Historic Building Grant. The total budgets for these grants are small, and neither is particularly intended to encourage renting.

New mechanisms to encourage private renting

Either way, central government will lack power to finance or implement its own policy. It is expecting to rely for that on just those bodies – the local authorities – that have been on the receiving end of the Department of the Environment's strict financial discipline over the past few years, and many of which are feeling understandably truculent. Without extra powers or inducements, English housing ministers may only be able to entreat, rather than direct, public investment in private or joint schemes. But their Scottish and Welsh colleagues can already do more (housing in Scotland and in Wales is the responsibility of the Scottish and Welsh offices respectively). Scottish Conservatives are still much more devoted to widespread state subsidy. This attracts much criticism in English Conservative circles, but it may help their housing policy. The Housing Corporation in Scotland and the (public) Scottish Special Housing Association combined on 1 April 1989 to create a new public body – Scottish Homes – which can, among other things, channel grant to private investors in rented housing. Housing in Wales was also set up in 1989 with similar functions.

The first substantial new organisation for developing and managing private rental housing since the second world war – Quality Street – was set up in 1987 by the former Director of Housing of Glasgow City Council. It hopes to attract public subsidy because it intends to cater for a range of incomes, including low-income tenants and former council tenants. It has attracted £600 million phased funding from the Nationwide Anglia Building Society. Its initial operations were rather restricted but it now owns more than 2,000 rented homes in various parts of Great Britain and has entered negotiations with local authorities over housing estates and new developments in England and Scotland.

Proposals for the further encouragement of long-term investment in private renting in England have been detailed elsewhere (Coleman, 1992; Best *et al.*, 1992). They include the creation of an English mechanism on the lines of Scottish Homes or Housing in Wales to channel public funds into approved private investment. The Housing Corporation itself, given a brief to encourage the private sector, might be given these powers to promote and subsidise approved private renting for average- or lower-income families. The private-sector bodies approved for this purpose would be able to compete even more effectively with housing associations if they could share the associations' privilege of being exempt from corporation tax. Some of the objections to subsidy could be avoided by concentrating such provision on particular areas. Inner-city sites are an obvious candidate, as part of the inner-city renewal programme. But rural areas also have particular problems in meeting the needs of low-income households, which are not yet adequately met by housing associations and where the subsidised private sector should play a role if more land were available for building, possibly under a covenant restricting the tenure at least for some period of time.

While accepting the inevitable initial pre-eminence of the housing associations as alternative landlords, there is a risk that arrangements made now, when there are few plausible alternatives, may shut out later alternatives for good. If low-rent housing has to be a long-term feature of the British housing scene, the government may be minimising the chances of genuine private-sector involvement in it by institutionalising association privileges, in part through the extra powers given to the Housing Corporation, and excluding competitors from them. In that case, the involvement of building societies and financial institutions will never be substantial, and the private companies committed to renting which are prominent abroad will never develop here. Subsidised renting in other countries often has mechanisms whereby individuals as well as institutions can invest for a safe if unspectacular return – as investing in renting always used to be. To encourage the same here would be a great opportunity to give the public a stake in the development of non-council renting, in the way that the building societies help their investment in ownership.

ALTERNATIVE ROUTES TO SUBSIDISING THE POOR

A high proportion of households will continue to need subsidy if they are to enjoy housing of an acceptable standard. Some – people with permanently low earning power – will need it all their lives. Others need it temporarily, while going through low-earning or high-cost parts of their life-cycle, or old age, or during periods of misfortune (e.g., unemployment, divorce with dependent children). Given the expense of such support and its likely effect on spending, saving, dependency and mobility, the different routes to providing subsidy for a given quality of housing provision have not yet been given the scrutiny they deserve.

Subsidy to a given standard of housing is likely – in theory – to have similar total costs whether given through housing benefit or through low rents. But in practice it may not. Where several systems of housing subsidy co-exist, it may be possible to evaluate the relative effectiveness of consumer subsidies and producer subsidies. Producer subsidies reduce rents below economic levels. They include the various subsidies which keep council and housing association rents sub-economic in the UK (and rents for public housing in the USA, such as Section 8 and Section 236 subsidies, which provided tax interest and depreciation allowances for private landlords), and tax and other subsidies for 'social housing' in former West Germany.

Consumer subsidies are transfers directed towards individuals to help them pay part or all of their rent, such as British Housing Benefit, US Housing Vouchers and *Wohngeld* in former West Germany (see Stahl and Struyk, 1985). In helping existing tenants in existing housing through different routes, the German and US experiences seem to suggest that individual housing benefit type support is more effective than low rents, which can increase total costs to well over market levels (Mayo, 1986). Initial results with the US Housing Voucher system, intended to replace the Section 8 scheme, appear to have been promising (Kennedy and Finkel, 1987) although critics, such as New York City (Stegman, 1987), insist that this scheme (which is cash limited, like Section 8) is provided on too small a scale. Encouragement for new building, however, may be more effectively done through front-end subsidy to landlords. Such subsidy is usually linked to low rents.

The government's own general aims of visible costs, rational choice and accurately targeted welfare might be better served, at least for existing tenants, through economic rents compensated by individual subsidy through a revised housing benefit. Poor households pay market prices for food, heating and transport, and the welfare system compensates them, to a greater or lesser degree. Expenditure figures from the Family Expenditure Survey show that, overall, housing accounted for a smaller proportion (16 per cent) of household expenditure than food (20 per cent) and only slightly more than transport and vehicles (15 per cent). Indeed the poorest third of families with children spend less on housing (9.7 per cent) than on alcohol and cigarettes (10.1 per cent), although this housing expenditure figure is substantially reduced from what it might otherwise be by low rents (Social Trends, 1987, table 6.11). In 1986, average weekly expenditure on housing per household was £55.10 for mortgage holders, £13.85 for council tenants, £19.10 for private (unfurnished) tenants and £31.93 for private (furnished) tenants (Social Trends, 1986, table 8.37).

The present policy takes for granted that a low non-economic rent regime should continue without assessing the direct and indirect costs of low rent or proposing alternative ways of housing subsidy. This means that the present hybrid system of producer and consumer subsidies will continue: sub-economic rents supported by rate payers and taxpayers and some private landlords' pockets, and also direct transfers to a majority of these households through Housing Benefit. This is likely to continue to inhibit movement between tenures and places.

The balance between low rent and individual benefit

The continuation of this system is inevitable for the time being. Proposals for radical reform would be politically very difficult. They would risk disturbing the vested interests, and the peace of mind, of millions of existing or prospective tenants. However generous the Housing Benefit compensation was made, the chief message of radical change would be the threat of higher rents. Within government there would be objections to a big increase in the Housing Benefit bill precisely because it is so visible compared to other routes of subsidy. Indeed, a substantial increase in Housing Benefit costs is already occurring following increases in council and housing association rents.

But the relative contribution to an individual's subsidy from low rents, and from Housing Benefit, are at present quite arbitrary. Neither is fixed or based on rational principles; each tends to be set by considering the level of the other. Housing Benefit levels have some effect on rent levels, or at least exert pressure to keep them low. For example, housing associations tend to oppose higher rents lest tenants just above Housing Benefit level be forced to pay too much of their incomes in rent. To spend more than 20 per cent of income on rent is regarded as undesirable by the National Federation of Housing Associations (1987), although the Housing Corporation (may or may not) have suggested in 1989 that a third would be reasonable (*Housing Associations Weekly*, 28 July 1989). At the same time the level of rents has been one consideration in setting the levels of Housing Benefit. 'Fair rents' themselves, which housing associations have defended so strongly, bear no relation either to the individual tenant's means or to the costs of providing the accommodation; neither, in many cases, do local authority rents (Audit Commission, 1986). In this system there seem to be no fixed points related to the cost of providing housing. When rents are related neither to the cost of provision nor to the means of each individual tenant then they are a poor form of subsidy. The Federal Republic of Germany withdrew federal support from low-rent schemes from 1986 precisely because it felt that they target help so badly, and concentrated its efforts on *Wohngeld* instead.

An alternative to the present system would be economic rents in all sectors. Housing benefit subject to regional ceilings determined by median rent levels, as in the US voucher system, would provide a complete variable subsidy to meet every tenant's unique and changing needs. There are two powerful objections to this (see Hills *et al.*, 1989). One arises from the 'poverty trap': that the formula whereby the benefit ceases to be paid as income rises may impose what are effectively very high marginal tax rates – possibly exceeding 100 per cent – on the additional earned income and thereby tend to deter employment and keep households dependent upon benefit. The 1988 reforms of housing benefit (see DHSS, 1985; Ward and Zebedee, 1988) are claimed to prevent negative tax rates partly through being based upon net, not gross, income, although the taper is steep and is blamed for imposing relatively heavy burdens upon lower-paid workers. The proverty trap should be avoidable. It is a consequence of the particular form of housing benefit used, not an inherent property of portable individual subsidy. It should be possible, for example, to delay the withdrawal of benefit over time, perhaps staged over a year, to make for an easier transition into employment or higher pay. The maximum amount payable could stop short of 100 per cent, perhaps at 90 or 95 per cent. Another serious objection is that housing benefit might cut off tenants from all signals about housing costs and minimise their perception of the value of their subsidy. This may lead of over-consumption and excess profits by landlords and uncontrolled increases in the subsidy bill. This may be answered by limiting benefit to rent levels around the median of the local distribution or rent levels, as in US practice, so that tenants wishing to rent at a higher level would have to pay the difference themselves.

The Housing Act does have some elaborate precautions against over-consumption, but they seem likely to curb only the extremes of over-consumption or of choice of expensive tenancies. The government has emphasised the importance of 'affordable' rents for council and housing association dwellings, which despite some suggestions that they might be up to 33 per cent of income (*Housing Associations Weekly*, 28 July 1989), will, according to the Housing Corporation circular of July 1989, be up to associations themselves to define according to local circumstances. They remain, therefore, completely undefined in relation to the cost of housing and services provided. For Housing Benefit purposes, rent officers define a 'reasonable rent' as a ceiling for benefit (but not the rent itself) and also ration the volume of housing eligible for benefit if claimants appear to be 'over-accommodated'. Benefit is not available for the full rent of particularly expensive properties, through another complicated procedure of rent assessment. The government has insisted that this is only likely to apply to a very small proportion of the accommodation in any area. Such caution may be necessary to avoid arousing unwarranted fears of general, high-rent increases amoung existing tenants. And it is politically difficult for the government to express bluntly the view – widely held – that many tenants in employment could well afford to pay substantially more for their housing than they do at present. The present government will not want to review housing benefit until more time has elapsed after the complex changes which were implemented in 1988. But it should be high on the agenda for future action.

Inevitable or not, this aspect of the policy may disconnect – for many claimant tenants – any link between the cost of their accommodation and their own means. This will matter more with higher new rents than it does now. Consequently it is already greatly increasing Housing Benefit costs (which are not cash limited), without Housing Benefit being designated as the main channel of subsidy. That might lead either to a demand for lowering of the ceiling of Housing Benefit after the event – which would be very unpopular – or for a reimposition of rent controls. The new market rents will apply only to new tenancies freely chosen by tenants. Existing tenancies of all kinds will continue much as before, with little scope for large rent increases although the example of new market rents may put upward pressure on 'fair rent' reassessments.

'Affordable', non-economic rents, like any non-portable subsidy, are likely to continue to obstruct labour migration and movement between housing sectors because people will be trapped by their subsidy. Subsidised rents always obscure real costs and distort expenditure and consumption of housing and other components of the household budget. This is why council accommodaiton makes a negligible contribution to labour mobility (Minford *et al.*, 1987; Salt, 1980, 1991) – and it may prevent the housing associations making much of a contribution there either. In theory, low rents should not discourage private-sector involvement in the take-over of public housing because such housing is to be valued for disposal on the basis of its net discounted rental flow. In practice, rents held down permanently by formula are not likely to be attractive to investors venturing into new territory. Low rents may tend to perpetuate our present two-nation system of housing, with low-rent estates only for poor people, with all the social problems which have been attributed to them.

DEPENDENCE ON HOUSING ASSOCIATIONS

One of the key elements of the government's rented housing policy is to encourage the devolution of local authority housing estates to other landlords, either at the instigatation of the local authority itself, using existing legislation, or at the request of the tenants under the new Tenants' Choice provisions of the Housing Act 1988. The Housing and Local

Government Act 1989 is meant to facilitate such transfers by imposing a more rational cost regime on local authority housing. If successful, this policy would have substantial social, economic and political consequences. But in practice the government seems to be depending almost completely on housing associations to deliver this policy.

There are good reasons for this. The housing associations have the enormous advantage of being already established. It is inevitable that they would be expected to play a key role in offering alternatives in rented housing. They are the only major source of management expertise other than local authorities (see Hills, 1987). They are the easiest answer to the question, 'Who will take on the local authority estates if tenants – or councils – wish to transfer?' The decayed traditional private rented sector is not a plausible response. The new model private rented sector, such as Quality Street, has little track record yet. The BES-based renting companies are obviously not appropriate.

Housing associations are also 'respectable'. They are free of the taint of Rachman, the invocation of whose name replaced critical thought in much of the debate on the Housing Bill. They claim to occupy the moral high ground, even if it is entirely at public expense. They concentrate on the lower paid and disadvantaged. Many operate in inner-city areas. A few have supported their renting through building for sale and through setting up subsidiaries to encourage low-cost home ownership and shared ownership. Some have also been keen to use the new opportunities of assured tenancy renting, under earlier legislation (Housing Act 1980, Housing and Planning Act 1986), to attract finance and extend their operations.

But on a closer look the housing associations may be less appropriate than they seem to fill this new role. Housing associations are part of the 'voluntary sector' and by inclination and funding are effectively part of the public sector, with which indeed they used to be classified. They do not, indeed cannot, make a profit. They have no shareholders in the conventional sense. They are not private, and yet cannot be privatised. In some important respects they do not compete with each other except for sites and public funding. Because they are handing out a publicly funded privilege they do not compete for customers – quite the reverse. Like local authorities, they run queues.

Under the present government, the market is meant to solve problems of choice, supply and demand. Accountability and the responsibilities of ownership are supposed to be important. It seems surprising that the government should be entrusting an important component of its housing policy to agencies, many of which rather like being within or on the edge of the public sector, which object to profits in housing and which have been protected by the taxpayer from the worry of raising private finance and by the rent officer from the trouble of setting their own rents.

Many housing associations – and the NFHA – have been hostile to some of the government's housing proposals (NFHA, 1987). It is understandable that they may not want to take over former local authority housing which would give them new problems but no new housing. Many enjoy a close connection with their local authority and depend on its goodwill for new projects. In Greenwich, for example, associations have agreed with the local authority to preserve the status quo (*Housing Associations Weekly*, 1988). The larger national associations are particularly vulnerable to local authority or opposition threats that they will not receive the co-operation they need over land, grants and planning if they play the government's game elsewhere in the country (*Voluntary Housing*, 1988). In some ways they look more like an extension of council housing, not a suitable replacement for it.

Housing association take-over would not necessarily diversify the social uniformity of estates which is regarded as one of its biggest problems. Housing association tenancy is also a one-class and one-quality operation. On their estates, housing associations can provide no staircase to better quality rented housing, although some are keen to help their tenants and others to owner-occupation through various low-cost home-ownership schemes, especially shared ownership. The movement has now targeted a narrower section of society than that housed by local authorities. On top of that, the Housing Corporation encourages the creation of ethnically segregated housing associations (37 by September 1989: see *Housing Associations Weekly*, 1989, and 66 by mid-1991) which effectively offer housing only to members of particular ethnic minorities in the wider British society. It seems astonishing that such restricted access should be legal.

At present, housing associations account for about 2.5 per cent of the housing stock. Local authorities account for 25 per cent. At its most radical the government might like to see all housing pass out of the direct control of local authorities. If housing associations were to take over all of it, then they would have to expand almost tenfold. Most people involved in housing associations might regard a doubling of their activity as a major and thought-provoking enterprise; but a tenfold increase would be seen as being out of the question. Admittedly this has not prevented some local housing associations from bidding for local authority housing which is more substantial than all their existing holdings. Whether they can succeed in managing such a rapid increase in their responsibilities remains to be seen.

No reforms of housing associations themselves are proposed to make their expansion easier. They remain impermeable to individual investment.

If they are to become much bigger, with bigger demands on their boards, then the voluntary status of those boards may no longer be adequate to attract further talented membership. There may be a case for permitting associations which wish to diversify their activities to privatise themselves: attract shareholder capital, make profits and issue dividends, attract corporate talent to run them by paying their boards. The building societies are a good example of a partly regulated housing institution which can channel the small investments of millions of people into ownership. Reformed housing associations could do something similar to encourage investment in rented housing.

Local authority transfers

The small scale of housing associations is one of the reasons why local authorities have started to create their own associations. Council motives in setting up these associations are mixed. Some councils would no doubt like to preserve their housing departments by turning them into housing associations. It could free them from constraints on capital receipts, from the continuation of right to buy, and from their new obligation to put out contracts to tender. It could help perpetuate contract compliance and continue the transfusion of taxpayers' money into direct labour organisations. In some areas, HA boards dominated by local councillors, together with suitable nominees from tenants' associations, could maintain political control. The DoE has issued firm guidelines to ensure independence of such housing associations from their parent local authorities. Housing associations will not be approved for transfers unless council board nominees are limited to a fifth or less. Councils cannot retain nomination rights. But the effective policing of all the new local authority-sponsored housing associations may be difficult if the idea catches on on a big scale.

More than 130 housing authorities out of 401 have expressed some interest in disposing of some or all of their housing under the mechanisms of the 1986 Housing and Planning Act. Most have been waiting to act until the result of the 1992 General Election was known. Some have responded to the new emphasis on alternatives to direct council ownership or management, and to the increasing financial pressures from central government against building mass council housing in the old style. Sixteen councils have so far transferred their stock, and two more have tenant approval to do so (as of November 1991).

Tenants are concerned about rents and about being handed over to private landlords (of which there is little chance, thanks to the Housing Corporation guidelines, which entirely favour housing associations). If there is a cliff of higher rents (whatever the Housing Benefit compensations) in front of a private transfer, few tenants will chooose to scale it. Nonetheless, an opinion poll (Gallup Poll, 1988) for the National Consumer Council revealed surprisingly favourable attitudes to the new choices. Of the respondents, 51 per cent preferred to stay with the local authority; 26 per cent did not know or needed more information; and equal proportions of the remainder would opt for a housing association, a building society or a co-operative. Given that very few will have any direct knowledge of housing associations, and none of building society renting (as none exists yet), these figures are rather remarkable. The 1 per cent who would choose a private company is not surprising in view of the poor reputation of existing private renting and the complete non-existence of modern systems of private renting which flourish overseas, for example in West Germany. So far, almost all transfer proposals have come from local authorities. Tenant-driven transfers may be much less likely to get off the ground until the results of the local authority initiatives are visible, and until the new housing finance proposals in the Local Government and Housing Act begin to have an effect upon council rent levels. One result seems to have been an acceleration of right-to-buy sales.

The Local Government and Housing Act 1989 is intended to make council rents respond more to the costs of providing housing by making local authority housing finance more tightly defined (DoE, 1988b and 1988c). From 1990 Housing Revenue Accounts (HRAs) have been ring-fenced to prevent their being subsidised from the rate fund account (or community charge). From 1990 a new Housing Revenue Account Subsidy provides housing subsidy from central government where necessary on national criteria (see DoE, 1990). Under the previous regime, 95 local authorities received additional subsidy from central government. The intention is to make local authority housing finance more efficient, to force councils to collect rents and reduce rent arrears and make use of empty dwellings, and to charge rents more commensurate with their housing costs (Audit Commission, 1986). One consequence will be to raise rents on average and reduce the advantage of remaining a council tenant over other forms of tenure. But many local authorities can charge very low rents without subsidy from central government or the rates. This is made easy because local authority housing is accounted for on an historic cost basis (see Hopwood and Tomkins, 1984), not on the basis of maintaining or replacing the property at current prices, as any commercially viable organisation would be obliged to do. Older housing is therefore on the books at the original nominal cost which appears to justify low nominal rent levels, irrespective of the costs of borrowing to maintain or replace them. The Local Government and Housing Act does not require local authorities to reform their acocunting methods. Instead the problem is being approached through a new method of determining rents. Furthermore, local authorities are obliged to pay part of local Housing Benefit payments out of any surplus which this system appears to generate.

Local authority rents are now determined through a formula by reference to the market value – not the cost – of dwellings through the valuation of dwellings sold to tenants or to others. Rent subsidy comes from one route, not three as in the past (main housing subsidy, rent rebate subsidy, part of RSG), by reference to a notional rent for each area determined yearly by the Secretary of State.

At a stroke the historic cost system of rent setting is being rendered nugatory, without the problem of abolishing it. Rents will come to reflect more than before (and more than new housing association rents) local differences in housing costs. This new local authority rent regime is a kind of regional policy without a regional policy: it may have a marginal effect of encouraging a drift to the north as the regional rent differential is likely to widen. Rents will not, of course, be market rents: the subsidy element prevents their rising to real market levels. In some cases there may even be a fall. The formula relates them to previous rents so increases are graded. In all respects this rent regime is quite different from that relating to new housing association assured tenancies. There, the target is 'affordability' (not precisely defined) by tenants just above the Housing Benefit level. The target rent will have considerable bearing on the form of finance adopted and on whether particular schemes can go ahead at all. Unlike the new housing association regime, the local authority changes apply to existing tenants as well as new tenants.

At April 1990, local authority rent increases (applying to all tenancies) ranged between 95p per week (below inflation) to £4.50. Over a third (156) of authorities were given the minimum guideline. The average increases were £1.15 in the north, £4.78 in the southeast. The new regime increased the average England rent of £21.51 per week in 1989 by 13 per cent to 1990 and by a further 15 per cent to reach £27.31 in 1991. The system explicitly avoids having a final target, based on affordability or anything else.

If tenants do wish to transfer, housing associations look like winning hands down because they enjoy some unique privileges. They can continue to charge sub-economic rents at the public expense. They too are run on an historic cost accounting basis. They do not need the option of selling empty houses, which in the DoE transfer rules require the permission of the Secretary of State (DoE, 1988e). If the private sector is to be allowed to compete in new developments, then government grant should also go to suitably approved elements in the private sector on the same basis, as the building societies and others keep saying (Melville-Ross, 1987). In respect of transfers, housing associations and private institutions are at least in theory on the same footing, as Housing Association Grant is only available for new development. But any institution the bulk of whose capital costs are paid from public funds and which are outside the Corporation Tax system, must be starting from a position of advantage. Unless the advantages between sectors are balanced, the only option for council tenants will be a transfer which remains effectively within the public sector.

LAND PRICES AND PLANNING CONTROLS

There is little sense in pointing to the merits of rented housing in helping labour migration and reducing unemployment when it is impossible to build houses – for rent or for sale – in some of the places where families and newly married couples want to live and where labour migrants want to go. Shortage of land in areas of high demand is the key to many problems in housing and to some extent in employment (CBI, 1988). A land price component of 40 per cent on average in Britain and much more in the southeast (Evans, 1987, 1992) contributes to the house price problem. It will tend to push up rents to the same degree. Shortage of land is at the root of disputes in rural areas between the claims of local people for lower-cost housing and the inflationary pressures exerted by newcomers. The solution is, in part, in the hands of local electorates themselves: to make more land available. But changes would need to be concerted, or driven from the top; otherwise the first area to move would be inundated with applications. Changes in planning powers for housing could encourage building for rent either through a new form of development permission solely for renting, or through the legal formalisation restrictive covenants on former public land – in rural areas as well as inner cities.

If this problem is not effectively addressed, then high price and relative shortage in the southeast and London will inhibit new rental development, or make it unduly expensive, just as it does with housing for owner-occupation. Some structure plans (see DoE, 1987) in the southeast were inadequate even for local household formation needs, never mind migration. The 1985- and 1989-based household projections suggesting faster than expected household growth – by an additional 200,000 in the south-east alone (DoE, 1988d, 1991) – give added urgency to this problem. SERPLAN (London and Southeast Regional Planning Conference) estimates have had to be revised upwards accordingly (Grigson, 1988) and other estimates based on notions of 'housing need' suggest even higher current and future shortfalls (Niner, 1989). At the end of the 1980s, government speeches suggested that there might be some cautious movement on central government planning policy and appeals, at least in areas outside the Green Belts. But by early 1992 it was apparent that the 'plan-led' policy implemented by the 1991 Planning and Corporation Act would create new restrictions on rural building development. The projected increase in housing demand arises primarily from new patterns of household formation (Dicks, 1988), especially the delay in marriage and the rise of divorce, age structure changes, longer survival and immigration from abroad, most of which goes to the southeast and London. It is not a consequence of people flooding into the southeast from the rest of Britain. There has been a small net migration out of the southeast to the rest of the country for some time, partly, it is believed, in response to the pressures created by planning constraints (see Muelbauer and Murphy, 1988).

In addressing this problem it is not good enough just to point to the remarkable achievement of getting almost half of land for housing in recent years out of vacant or re-used urban land. That is fine if the land is won back from post-industrial dereliction, but less good if it is merely pushing up urban densities by building on back gardens. About a quarter of the land used for housing in the southeast in 1987 – 1,300 acres – came from surburban back gardens. Increasing surburban densities, and the destruction of surburban open space, is rapidly emerging as a major environmental problem – with a bigger electorate to be concerned than can be marshalled by the rural areas. Not all people want to live in cities, and some urban centres might still benefit by giving some of their residents a chance to live elsewhere – many will certainly want to do so if the government's plans to widen choice in housing actually work.

CONCLUSIONS

The Housing Act 1988 and the Local Government and Housing Act 1989, together with other measures such as BES, have together satisfied many of the requirements for a successful reform of renting. But in the private sector, if there is long-term improvement in the financial competitiveness of renting, then growth in new investment for rent will be confined to a limited market. It would also be a useful achievement even to slow or stop the decline in the small traditional landlord, who is an important and even a majority provider of rented housing in most countries' rental systems.

The Business Expansion Scheme has already succeeded in attracting substantial new investment into private rented housing for the first time since the 1930s. In doing so it may change public perceptions about the whole private rented sector. But the minor proposals to encourage private renting put forward by the Conservatives at the 1992 General Election provide no evidence yet that the government is considering further measures to attract private-sector institutions into long-term renting. With them, the transformation of the rented scene could be radical and permanent. New rented housing could be a modern and attractively marketed product, very different from the 'traditional' renting to which the Labour Party has in the past been automatically opposed and concerning which public fears are so easy to excite.

Outside housing, economic and social policy is based on notions of the benefits to the individual arising from competition and efficiency, the primacy of consumers and their ability to choose with fewer restrictions and regulations. The 1989 Local Government and Housing Act has imposed imaginative and radical reforms on local authority housing finance. For the first time, choices in the provision and consumption of such housing will reflect local costs, however buffered by the protection of subsidy. But in the housing association sector, ministers may be in danger of justifying their housing policy not so much on the grounds that it will work, but rather that it cannot do any harm, because it is safely wrapped in a system of approvals, regulation and controls and will not, in the hands of the housing associations, really be in the private sector at all.

REFERENCES

ADC (Association of District Councils) (1989) *Meeting Housing Needs: A Report on Research for the Association*, Glen Bromley, London: ADC.

Audit Commission (1986) *Managing the Crisis in Council Housing,* London: HMSO.

Best, R., Kemp, P., Coleman, D., Merrett, S. and Crook, T. (1992) *The Future of Private Renting: Consensus and Action*, York: Joseph Rowntree Foundation.

Boléat, M. (1989) *Housing in Britain*, London: Building Societies Association.

Bover, O., Muelbauer, J. and Murphy, A. (1988) *Housing, Wages and UK Labour Markets*, Discussion Paper Series No. 268, London: Centre for Economic Policy and Research.

Burnett, J. (1986) *A Social History of Housing 1815–1985*, 2nd edition, London: Methuen.

Coleman, A. (1985) *Utopia on Trial*, London: Hilary Shipman.

Coleman, D.A. (1992) '1992: A Real Opportunity for Private Rented Housing?' in *The State of the Economy 1992*, pp. 65–80, London: Institute of Economic Affairs.

Coleman, D.A. and Bloch, A.C.D. (1988) *New Opportunities for Renting – a Builder's Guide*, London: Housebuilders Federation.

Confederation of British Industry (CBI) (1988) Companies and the Housing Market, *CBI News*, March 1988.

Crook, T., Kemp, P., Anderson, I. and Bowerman, S. (1991) *The Business Expansion Scheme and Rented Housing*, York: Joseph Rowntree Foundation.

Daunton, M.J. (1987) *A Property-Owning Democracy? Housing in Britain*, Historical Handbooks, London: Faber & Faber.

Department of the Environment (DoE) (1985) *An Inquiry into the Condition of the Local Authority Housing Stock in England*, London: DoE.

——(1987) *Land for Housing Progress Report*, London: DoE.

——(1988a) *English House Condition Survey 1986*, London: HMSO.

——(1988b) *Local Government in England and Wales: Capital Expenditure and Finance*, consultation paper, London: DoE.

——(1988c) *New Financial Regime for Local Authority Housing in England and Wales*, consultation paper, London: DoE.

——(1988d) *1985 Based Estimates of Numbers of Households in England, the Regions, Counties, Metropolitan Districts and London Boroughs 1985–2001*, London: DoE.

——(1988e) *Large Scale Voluntary Transfers of Local Authority Housing to Private Bodies*, London: DoE.

——(1990) *Housing Subsidy and Accounting Manual*, London: HMSO.

——(1991) *Household Projections in England 1989–2011; 1989-based Estimates of the Numbers of Households for Regions, Counties, Metropolitan Districts and London Boroughs*, London: HMSO.

Department of Health and Social Security (DHSS) (1985) *Housing Benefit Review: Report of the Review Team*, Cmnd 9520, London: HMSO.

Dicks, M.J. (1988) *The Demographics of Housing Demand: Household Formation and the Growth of Owner-occupation*, Bank of England discussion papers No. 32, London: Bank of England.

Evans, A.W. (1987) *House Prices and Land Prices in the South East – a review*, London: Housebuilders Federation.

——(1988) *No Room! No Room!. The Costs of the British Town and Country Planning System*, Occasional Paper 79, London: Institute of Economic Affairs.

——(1992) 'Town Planning and the Supply of Housing', in *The State of the Economy 1992*, London: Institute of Economic Affairs, pp. 81–93.

Gallup Poll (1988) *Omnibus Report. Council Tenants 30 March–12 April 1988*, London: Social Surveys (Gallup Poll) Ltd.

Greater London Council (GLC) (1986) *Private Tenants in London*, The GLC Survey 1983–84, London: GLC.

Grigson, W.S. (1988) *Housing Provision in the South East. Report to SERPLAN on a revised distribution for the 1990s* (see also SERPLAN's Statement RPC 1350 with further revisions 1989), London: SERPLAN.

Halifax Building Society (1987) *The Housing Bill 1987*, Halifax: Halifax Building Society.

Hallett, G. (1988) *Land and Housing Policies in Europe and the USA: A Comparative Analysis*, London: Routledge.

Hills, J. (1987) *The Voluntary Sector in British Housing: the Role of British Housing Associations*, Welfare State Programme Discussion Paper No. 20, London: STICERD, London School of Economics.

Hills, J., Berthoud, R. and Kemp, P. (1989) *The Future of Housing Allowances*, London: Policy Studies Institute.

HRH The Duke of Edinburgh (1985) *Inquiry into British Housing: Report and Evidence (two volumes)*, London: NFHA.

Holmans, A.E. (1987) *Housing Policy in Britain: A History*, London: Croom Helm.

Hopwood, A. and Tomkins, C. (1984) *Issues in Public Sector Accountancy*, Deddington, Oxford: Philip Allan.

House of Commons (1982) *The Private Rented Housing Sector*, volumes 1 to 3, HC40, House of Commons Environment Committee, London: HMSO.

Housing Corporation (1988) *Tenants Guarantee. Guidance for Registered Housing Associations on Housing Management Practice*, London: Housing Corporation.

Inland Revenue (1987) *The Business Expansion Scheme*, Leaflet IR51, London: HMSO (at any tax office).

——(1988) *Business Expansion Scheme Press Release*, 14 April 1988.

International Monetary Fund (1989) *World Economic Outlook*, Washington, DC: International Monetary Fund.

Kennedy, S.D. and Finkel, M. (1987) *Report of First Year Findings for the Freestanding Housing Voucher Demonstration*, Washington, DC: US Dept of Housing and Urban Development.

Labour Party (1989) *Meet the Challenge, Make the Change*, London: Labour Party.

Lakhani, B. and Reed, J. (1987) *National Welfare Benefits Handbook*, 17th edition 1987–88, London: Child Poverty Action Group.

Leeds Residential Property Association (1987) 'A New Start for the Private Rented Sector', Leeds: Leeds Residential Property Association.

Mayo, S.K. (1986) *Sources of Inefficiency in Subsidised Housing Programs: A Comparison of US and German Experience*, Journal of Urban Economics, vol. 20, pp. 229–49.

Melville-Ross, T. (1987) 'Paying for Housing Policy', speech to London School of Economics, 6 October 1987, London: Nationwide Anglia Building Society.

Minford, P., Peel, M. and Ashton, P. (1987) *The Housing Morass. Regulation, Immobility and Unemployment*, Hobart Paperback 25, London: Institute of Economic Affairs.

Muellbauer, J. and Murphy, A. (1988) *House Prices and Migration: Economic and Investment Implications*, London: Sholander, Shearson, Lehman and Hutton Securities Research Report.

National Association of Home Builders (1986) *Home Building after Tax Reform: a Builders Guide*, Washington, DC: National Association of Home Builders.

National Federation of Housing Associations (NFHA) (1987) *Rents, Risks, Rights. NFHA Response to the Government's Proposals*, London: NFHA.

Niner P. (1989) *Housing Needs in the 1990s. An Interim Assessment*, London: National Federation of Housing Associations.

OPCS (Office of Population Censuses and Surveys (1991) *The 1991 Private Renters' Survey: Preliminary Results*. OPCS Monitor SS 91/2, London: OPCS.

Power, A. (1988) *Council Housing: Conflict, Change and Decision Making*, Welfare State Programme Discussion Paper No. 27: London: STICERD, London School of Economics.

Salt, J. (1980) 'Labour Migration, Housing and the Labour Market in the Inner City', in A. Evans and D. Eversley (eds) *Employment and Industry*, London: Heinemann.

——(1991) 'Labour Migration and Housing in the UK: an overview', in Allen, J. and Hamnett, C. (eds) *Housing and Labour Markets*, London: Unwin Hyman.

Stahl, K. and Struyk, R.J. (1985) *US and West German Housing Markets*, Washington, DC: The Urban Institute.

Stegman, M.A. (1987) *Housing and Vacancy Report. New York City 1987*, New York: City of New York Department of Housing Preservation and Development.

Todd, J., Bone, M. and Noble, I. (1982) *The Privately Rented Sector in 1978*, London: HMSO.

Todd, J.E. (1986) *Recent Private Lettings 1982–84*, London: HMSO.

Ward, M. and Zebedee, J. (1988) *Guide to Housing Benefit*, London: Institute of Housing.

Whitehead, C.M.E. and Kleinman, M. (1987) 'Private Renting in London: Is It So Different?', *Journal of Social Policy*, vol. 16, no. 3, pp. 319–48.

7 Issues of race and gender facing housing policy

Norman Ginsburg and Sophie Watson

Race and gender issues in housing are still often seen as marginal to the 'real' questions and are at best tacked onto the end of any discussion or debate, rather than integrated into mainstream analysis. Until the day comes when this is no longer the case, there remains a necessity to elaborate both a feminist and an anti-racist perspective in a collection such as this, in recognition that many of the points have been reiterated over several years with often limited impact (see, for example, Austerberry and Watson, 1981; Watson with Austerberry, 1986; Jacobs, 1985; Karn, 1983). It is also true to say that gender and race issues can be over-emphasised as issues necessarily specific to women and to black people,[1] when often very similar points can be made about all low-income households in a highly privatised housing market such as exists in Britain today. This chapter will not attempt to draw simplistic parallels between the housing situation of women and of black people in Britain. We shall merely present an account of each, and examine the prospects for each in the 1990s.

BLACK PEOPLE AND HOUSING POLICIES

An index of governments' unwillingness to tackle the housing situation of black people in Britain is their failure to monitor housing conditions and housing stress specifically in relation to Britain's ethnic minorities. This in itself helps to create the impression that there is no real problem. Of course, there is much evidence to the contrary. The only national surveys of the housing conditions of different ethnic groups have been done by the Policy Studies Institute, most recently in 1982 (published in Brown, 1984). In terms of 'basic amenities' (exclusive use of hot and cold water, inside toilet, etc.), there was a dramatic improvement for the whole population between 1974 and 1982 and a sizeable narrowing of the gap between blacks and whites. In terms of persons per room, however, the gap between blacks and whites seems to have grown over the same period. The same is true for other amenities: in 1982 Afro-Caribbeans were three times and Asians two times more likely than whites not to have a garden. More recent data on London for 1985 suggests that 19 per cent of Afro-Caribbeans and 29 per cent of Asians in the capital are 'in housing need' compared to 12 per cent of whites, using an index of physical conditions including persons per room (London Research Centre, 1989, table 3.13). The government's 1986 English House Condition Survey (DoE, 1988) differentiates households only by the birthplace of the head of the household, thus putting the majority of black Britons in the UK category. Of dwellings where the head of household was born in the UK, 3.8 per cent were unfit and 12.5 per cent required 'urgent repairs estimated to cost more than £1,000', compared to 9.3 per cent and 21.9 per cent respectively where the head of household was born in the Caribbean or Asian sub-continent (DoE, 1988, table 6.2). Since the last such survey in 1981 there has been little change in the proportions of dwellings which are either unfit or in poor repair. Although the evidence is inadequate in many respects, we may safely conclude that relative inequalities in housing conditions between blacks and whites remain substantial, and in some respects have increased, a trend likely to continue into the 1990s if present policies prevail.

A second key aspect of housing and race in Britain is the growth of official homelessness over the past decade, which has differentially affected black people, while their rights and their treatment under the Homeless Persons Act[2] have been inferior. According to the London Housing Forum:

> In 1985/6 around 32 per cent to 40 per cent of households accepted as homeless by local authorities in London were Black though they comprised 10 per cent of households. Black households are therefore three to four times as likely to become homeless as white. In some boroughs the figures are particularly high: Lambeth 45 per cent; Hackney 49 per cent; Brent 75 per cent; Tower Hamlets 90 per cent. As a consequence Black people form a high proportion of those forced to endure the appalling conditions in bed and breakfast.
>
> (London Housing Forum, 1988, p. 19)

This critical situation is illuminated further by a 1986 survey in Brent (see Bonnerjea and Lawton, 1987), where black people are about 50 per cent of the population and 70 per cent of the borough's official homeless. The number of Afro-Caribbean homeless is twice that of the Asians and three times that of the whites in Brent. The burden of homelessness is falling increasingly on black women: 46 per cent of Brent's official homeless were single mothers in the Brent survey. Bonnerjea and Lawton (1987, p. 65) found 'not so much that black families were treated less favourably than average, but that white families were sometimes treated more favourably than average'. Thus for example, white families tended to be in better temporary accommodation in hostels and hotels. Brent, like almost all local authorities, does not effectively monitor its policy practices on such criteria.

Processes of more overt discrimination against homeless Bangladeshis in Tower Hamlets emerge from a Commission for Racial Equality study covering the period 1984–85 (see CRE, 1988a). The CRE study found 'significant differences' (CRE, 1988a, p. 10) in the time spent in temporary accommodation by Bangladeshi families compared to whites. This was not explained, as the council claimed, by differences in family structure between the two groups. There was also evidence that Bangladeshis were offered inferior temporary accommodation – for example, more than thirty miles distant in Southend. The Home Affairs Committee of the House of Commons noted in a report in 1987 that for the homeless Bangladeshis in Tower Hamlets 'conditions in bed and breakfast hotels were often appalling, with severe overcrowding, lack of basic amenities . . . insect infestation and fire and safety hazards' (Home Affairs Committee, 1987, p. ix). The CRE concluded that Tower Hamlets had contravened the Race Relations Act in four respects, and a non-discrimination notice was issued against the council in November 1987 with an extensive range of policy recommendations. It is not clear to what extent these are being implemented effectively, but in April 1987 Tower Hamlets adopted a much tougher policy on intentional homelessness. This included the cancellation of hotel bookings for ninety homeless applicants, most of whom were Bangladeshis, whose families were waiting in Bangladesh for entry into the UK. As the CRE explain, 'the council declared them intentionally homeless on the grounds that they had accommodation in Bangladesh which was "available for occupation" and which it was "reasonable to continue to occupy"' (CRE, 1988a, p. 55). An appeal against this decision was finally rejected by the Court of Appeal in April 1988, despite the fact that the appellants had lived and worked in the UK for between 16 and 28 years. The CRE conclude that

> the outcome of the Tower Hamlets cases is clearly very important and will have a significant impact on the housing opportunities of Bangladeshi separated families who, because of their pattern of immigration and settlement, are more likely to be declared intentionally homeless.

> (CRE, 1988a, p. 56)

Indeed this decision will affect all such families, most of whom are black. This is one key example of a wider process of tightening eligibility for accommodation and broadening the definition of intentional homelessness under the Homeless Persons Act, which is discussed further here in relation to independent women. It is a process which will differentially affect black people in the 1990s.

Residential segregation and institutional racism

Up to the mid-1970s the exclusion of most black people from council housing was very significant in shaping residential segregation and tenure patterns. State-sponsored neglect of the private rented sector and low-income owner-occupation left black communities to organise themselves within the market by self-help and self-organisation, largely within poorer residential areas of the 'inner city' vacated by whites. The processes of residential segregation have continued since then with local and national policies playing a more proactive role. Since the early 1970s black people have secured council tenancies, but have been differentially allocated inferior accommodation. This has been brought about essentially by forms of institutional racism in the provision and allocation of council housing,[3] according to evidence from studies in Birmingham by Henderson and Karn (1987), in Tower Hamlets by Phillips (1986), in Liverpool by the CRE (1989), in Nottingham by Simpson (1981) and in Hackney by the CRE (1984). The three forms of institutional racism involved are:

1　Relative inadequacy in the physical standards of dwellings in relation to black applicants' needs, particularly the numbers of bedrooms, due to a failure to take into specific account the needs of black applicants in the house construction and acquisition programmes of the past;
2　Formal local policies creating differential access for black applicants such as dispersal policies, residency requirements, and exclusion of owner-occupiers, co-habitees, joint families and single people from housing waiting lists;
3　Managerial landlordism involving racialised assessment of the respectability and deserving status of applicants, assumptions about preferred areas of residence for different racial groups and about the threat of racial harassment by whites causing avoidable trouble for managers.

None of these forms of institutional racism depends on explicit, direct discrimination by staff and politicians. Such attitudes are, of course, rarely expressed publicly; they are more likely to exert their influence quietly and routinely. They are not, however, essential to the process, though of course they can and do contribute to it.

Beyond these processes, account must also be taken of what can be described as structural racism. This has several aspects in relation to council housing. First, public expenditure on housing has suffered greater cuts in real spending than any other public expenditure programme over the years since 1976 in a period when black pressures for such spending have become more insistent. Second, almost all of the council houses sold since the passage of the Housing Act 1980, well over one million homes, have been on the more attractive, suburban estates. This has hugely and differentially benefited white tenants, the 'respectable' working-class people who had benefited from racialised allocation in previous decades. Here institutional and structural racism intersect strongly. Black tenants' chances of transferring to the suburbs and thence taking full advantage of buying a council house have been cut down substantially by the combined effects of these national and local policies. The restraint in spending on council house building, rehabilitation and maintenance, allied with council house sales, have solidified and extended the racial inequalities already existing in the mid-1970s. These effects seem likely to continue into the 1990s under present policies.

Government support for home ownership and the house price booms of the 1970s and 1980s have had an ambiguous impact on the black communities. Certainly some black people, particularly in London, have been able to capitalise on their presence in the owner-occupied market, though it is impossible to quantify this in relation to similar benefits accruing to white owner-occupiers. Outside London, properties owned by black people, as Luthera says,

> sustain their value with difficulty due to overrepresentation of blacks in declining areas and because they find it difficult to repair properties due to mounting unemployment. Asians, having purchased their property without benefit of tax relief, now find these just sustaining their value in areas of high unemployment.
>
> (Luthera, 1988, p. 131)

Mortgage tax relief has been the most significant form of government financial assistance for housing in the 1980s, but it is of course extremely regressive, benefiting relatively little those on low incomes like many black people. In addition, as Karn *et al.* (1985) have documented, many black owner- occupiers bought their properties with informal, high-interest, short-term loans which did not attract mortgage interest tax relief.

Residential segregation in the owner-occupied market is sustained by the everyday activities of lenders, estate agents, surveyors, solicitors and vendors. These institutionalised and informal processes are not unlike those which operate in the allocation of council housing, but they are rarely prosecutable or accessible to research. Robin Ward has suggested that for estate agents, the process of discriminating between different categories of potential house-buyer is a central feature of successful estate agent practice. To estate agents the distinction between racial steering (which is discriminatory) and disposing of property in a way calculated to give the highest return and lead to the maximum flow of commissions may be a narrow one.

(Ward, 1982, p. 14)

It is quite difficult to detect and prosecute such forms of private institutional racism. The CRE have occasionally taken action against explicit racial differentiation by estate agents. An example of this was the case of Richard Barclay & Co. in Clapham, a firm which was consistently and explicitly discriminating against Asian buyers (see CRE, 1988b).

Valerie Karn has also described how institutional racism operates in the finance of owner-occupation by the banks, insurance companies and building societies. She suggests that it takes three forms (Karn, 1983, p. 178):

1 Abnormally bad treatment of black buyers and black areas – i.e., high refusal rates for black applicants and for sales in black areas;
2 Abnormally good treatment of white buyers and white areas – e.g., willingness to lend to white non-savers;
3 Stereotyping of preferences sustaining the expectation that blacks will only want to buy in areas where their community is concentrated and that no whites will want to buy there or remain living there.

The intersection of local institutionally racist processes and socio-structural factors here is complex and not fully researched. Home loan institutions can argue that black applicants are likely to be poorer and the houses they want to buy more likely to be in a poor condition, due to the position of black people in the socio-economic structure. Valerie Karn was writing the above in the 1970s. It is not clear whether and to what extent the liberalisation of the home loans market in the 1980s has undermined some of these processes. In the 1990s lending institutions will hopefully come under renewed pressure to monitor and reform their practices in relation to ethnic applicants.

Racial Harassment [4]

Racial inequalities in housing, particularly in the rented sectors, cannot be explained simply in terms of the effects of the class structure combined with the structural concentration of black people in lower-income groups. The sustenance for racialised allocation of rented housing is not simply provided by national and local government housing policies. Nor can it be explained solely in terms of either directly prejudicial or institutionally racist processes within local authorities and housing associations, important as these are. From the studies of racialised allocation of council housing mentioned above, it becomes quite clear that racial violence and harassment in and around the home are very important factors sustaining racial inequalities in housing. The fear of possible anti-black racial violence is a factor that often motivates housing managers and black prospective tenants in allocating and choosing tenancies. Racial attacks, racial harassment and the threat of them are thus very significant in perpetuating racism and racial inequalities in housing in contemporary Britain.

Until recently at least, racial harassment and policy to combat it have not really been considered to be a legitimate part of housing policy, judging by the academic and professional literature. Housing policy analysts and housing managers sometimes seem unaware of the effects of racial harassment or see it as part of the social context beyond their realm, as a policing or legal matter. However, if racial harassment and racial violence are important influences on the policy process, then effective policies and practices to counteract them must surely be an essential part of progressive housing policy for the 1990s. For many years now there has been considerable hard evidence on the extent of racial harassment and its effect on housing inequalities, and a long list of policy recommendations to counteract it (see Bonnerjea and Lawton, 1988); Home Office, 1981; Racial Attacks Group, 1989). Unfortunately, so far there has been no published, systematic monitoring of the implementation of such policies, few legal cases against perpetrators and apparently limited preventive work by welfare agencies. Recent critical commentaries on the question of racial harassment and housing such as the CRE report *Living in Terror* (1987) have put most of their emphasis on local authority action. The police and policing policy are often so remote and unaccountable that they are too difficult to tackle directly at the grass roots level, except in particularly scandalous cases.

The CRE give a list of eleven 'key components of a racial harassment policy', the most important being monitoring of all incidents, clear designation of officer responsibilities and specialist staff, racial harassment as grounds for eviction and high-priority transfer for victims where necessary. Of those local authorities who had such policies, fewer than a third had reviewed them and fewer than half were monitoring them. The conclusion must be, therefore, that except in a few cases, serious preventive policies either do not exist even on paper or, where they do exist, they are not being implemented properly. In regard to housing departments, most discussion has focused on transfer of victims and eviction of perpetrators.

The former is clearly the norm and the latter extremely rare. The CRE report (1987) says that the number of victim transfers in their survey was 'significant', though precise figures were often unavailable from local authorities. Without action against perpetrators, victim transfers, or 'management transfers' as they are officially called, are urgent and vital but they seal a victory for the perpetrators. Victim transfer is disastrous as a long-term solution, yet it is being widely implemented.

There is an impressive array of legal powers which, in theory at least, local authorities, the police, the CRE and victims themselves can call upon to take action against perpetrators. Duncan Forbes (1988) has written a lengthy manual on the various legal remedies available to local authorities in tackling racial harassment. This book ought to have a considerable impact on the legal departments of local authorities who have been rather hesitant about taking up cases. The only cases of any significance taken up by local authorities so far have used Schedule 2 of the 1985 Housing Act. This sanctions eviction if 'the tenant or person residing in the dwelling house has been guilty of conduct which is a nuisance or annoyance to neighbours' (CRE, 1987, p. 26). This is called the 'nuisance clause'. Some local authorities have amended their tenancy agreements to include racial harassment as grounds for eviction, though there are apparently 'no reported cases of eviction proceedings being processed for breach of such a clause in an agreement' (MacEwen, 1988, p. 63). So far there have only been a handful of 'nuisance clause' cases, not all of them successful. Beyond victim support and action against perpetrators what is also required is more indirect public support for community and voluntary initiatives. This already happens to some extent, of course, but more activity is required in terms of anti-racist communication and dialogue with tenants. This means challenging white attitudes condoning racial harassment in ways which constructively change confused, individual prejudices (see Dame Colet House, no date). It also means ensuring that paper policies are communicated positively to all tenants.

Black people and the Housing Act 1988[5]

The Act differentially affects black people in general because it concentrates on lower-income households, amongst whom blacks are strongly represented. In particular, the majority of Afro-Caribbeans (60 per cent in 1982) are tenants, as are many Asians (28 per cent in 1982; see Brown, 1984, table 29). It is possible that more black people would move into rented housing from owner-occupation if secure, affordable and adequate accommodation were available, because the housing conditions of black owner-occupiers are frequently poor (see Brown, 1984, table 35).

Afro-Caribbeans are four times more likely to be housing association tenants than whites and Asians, so they are particularly affected by the changes in finance and organisation of housing associations discussed in previous chapters.

During the Parliamentary passage of the 1988 Housing Act, the CRE and the opposition supported amendments to alter the 'colour blindness' of the legislation. This succeeded in bringing the Housing Corporation and Housing for Wales (the new body overseeing housing associations in Wales) under the sway of Section 71 of the Race Relations Act 1976. This makes it a duty of every local authority 'to eliminate unlawful racial discrimination and to promote equality of opportunity and good relations between persons of different racial groups'. This will not, however, apply to individual housing associations, 'approved landlords' or Housing Action Trusts (HATs) because, according to the then housing minister, William Waldegrave, 'we do not believe that Section 71. . .can be suitably applied to the specialist providers of housing with their narrower function' (CRE, 1988c, p. 2). Section 71 legitimises local authority equal opportunities and anti-racist policies, which it might be suggested have not thus far had a dramatic effect on local authority housing practices, (see, for example, Phillips, 1987). The CRE is soon to publish a Code of Practice on equal opportunities policy and practice for local housing authorities and the Housing Corporation. As Phillips says, 'the National Federation of Housing Associations has . . . provided guidance and advice to its members [on equal opportunities], but individual associations have been slow to implement its proposals' (Phillips, 1987, p. 113). We must wait to see how energetically the Housing Corporation pursues its new duties under Section 71.

As noted above, housing associations have been a particularly important provider of rented housing for black people. Forman (1988) suggests that black and other ethnic minority housing co-operatives have been one positive solution to the black housing crisis, describing how the Bangladeshi co-ops were established in Spitalfields out of the 1970s' struggle against City property developers and the Greater London Council. By the late 1980s there were six main housing co-ops there, housing 2,000 people. According to Phillips, 'of approximately 2,600 associations registered with the Housing Corporation in 1987, only about 20 are controlled by black people and less than half of this number are black owned' (Phillips, 1987, p. 116). The Federation of Black Housing Organisations believes that there are about a hundred black and ethnic minority housing associations, most of them not registered, not publicly funded and very small. All the registered black housing associations are small, the largest being Presentation in South London with 600 homes. According to Grant, registration depends on a track record and track record is impossible unless the group is registered and looks viable – the classic Catch-22. But extensive lobbying by black and ethnic minority housing campaigners has eventually filtered through to the higher authorities. In 1986 the Housing Corporation grandly announced its five-year programme. The aim was to register five new black and ethnic minority housing associations per year and provide capital funding.

(Grant, 1988, p. 30)

This is being fulfilled, but the new housing associations are being squeezed for cash with which to train staff and to develop legal, financial, counselling and translation skills. The 1988 Housing Act is going to make small housing associations much less viable as independent entities because of the pressure to raise private finance.

Black tenants in the private rented sector are already in an adverse position compared to whites. The Greater London Council (GLC) survey of private tenants in London (GLC, 1986) found that only 42 per cent of 'non-white' households had security of tenure, compared with 66 per cent of whites. The 1988 Act does not, of course, deal with the informal private rented sector outside the Rent Acts upon which most black private tenants depend. Direct discrimination, informally exercised, remains widespread amongst small private landlords. The GLC survey found that 22 per cent of black households had suffered serious landlord harassment, compared to 13 per cent of whites. The 1988 Act does provide for better financial compensation from the courts for such harassment as described above, but amendments to strengthen the law on *racial* harassment of tenants were rejected during Parliamentary debates on the 1988 Act. An amendment was accepted which provides that people who are transferred within a tenure for any 'management reason' (code phrase for racial or sexual harassment) will enjoy the same rights that they enjoyed in the previous property. A thorough anti-racist policy programme for the private rented sector has been published by London Against Racism in Housing (1988).

The effects of the further privatisation of council housing on black people will be mixed but on the whole adverse. The implications of Tenants' Choice for black tenants is ambiguous. On the one hand if it leads to a switch from council tenancy to housing co-operatives and other kinds of housing association, black people may have a lot to gain if those organisations provide a better service and are committed to anti-racist practices. On the other hand, if housing associations and approved landlords are not so committed, it could be worse than having the local authority as landlord. Tenants' Choice may also be used by white tenants to opt out of council control partly in order to keep out black tenants. As Errol Lawrence has said,

> a key determinant . . . will be the level of organisation amongst black
> tenants on council estates. On a few estates black tenants are very well
> organised (e.g. Broadwater Farm, Stonebridge, Clapton Park) and are in
> a good position both to assess the implications of 'opting out' and if
> necessary to resist the privatisation of their estate . . . . [O]n most estates,
> however, the organisational frameworks that would enable black tenants
> as a group to make informed choices about their housing tend not to be
> very well developed.
>
> (Lawrence, 1987, p. 6)

In these latter situations, black tenants are likely to be vulnerable to being
squeezed out by predator landlords, high rents, harassment and so on. Yet
the successful campaign against HATs led in part by black council tenants
from Brixton and Peckham shows that government expectations of black
disorganisation can prove very wide of the mark. On balance, then, the
housing outlook for black people in Britain is likely to be worsened by the
1988 Act and its accompanying policies (see also London Housing Unit,
1988, section 2), though it may have a positive impact for a substantial
minority.

GENDER ISSUES IN HOUSING

Gender in housing issues can be analysed at two main levels: first, the ways
in which housing reinforces dominant power relations, specifically those
between the sexes; and second, the ways in which housing reinforces
women's dependence and lack of autonomy in society generally. Both of
these often articulate with racialised policy processes, placing an acute
burden on black women (discussed later). In relation to women's lack of
autonomy, the major question is the access to housing that women have in
their own right – that is, not as a dependants of men. The dominant form of
housing tenure in Britain is, of course, owner-occupation, with a doubling
of the number of owner-occupied dwellings between 1961 and 1987, from
just over 7 million to 14.5 million. The increase has been particularly
marked over the past decade as a result of the deliberate privatisation
strategy pursued by the Thatcher governments. Thus by far the biggest form
of public financial assistance for housing is mortgage tax relief for
home-owners. Such an emphasis on one tenure at the expense of others is
discriminatory towards women, as will be discussed below.

The general statistics on home-ownership levels obscure very real
differences between household groups in their access to this tenure. The
highest levels of home ownership exist amongst married couples, with
particularly high levels of mortgaged ownership. Thus approximately 70

per cent of married couples are owners or purchasers of their homes, compared to less than 50 per cent of single householders (see DoE, 1986). When the figures are disaggregated by sex, the discrepancies are particularly marked. More single men than single women have mortgages, particularly in the Greater London area where 25 per cent of the loans distributed by building societies in 1986 went to single men, compared to 14 per cent to single women. Similarly the considerable impact of divorce and separation on women is reflected in lower proportions of mortgagees amongst divorced women.

These important differences between men's and women's access to what is generally the most privileged tenure in Britain can be explained primarily in economic terms, although doubtless several other factors, such as discrimination (see EOC, 1978), need to be taken into account also. The ratio of house prices to average earnings in 1987 stood at 3.8 per cent. If this statistic is considered alongside the staggering increases in house prices throughout the country, it is obvious that lower-income households are liable to be excluded. Despite the changes in the levels of women's participation in the workforce, and some apparent shifts in women's status in this society, women still take primary responsibility for domestic and child-rearing work and face a sex-segmented and hierarchical labour market. Women are concentrated in the lower-paid occupations and at the bottom end of the hierarchies. On average, therefore, women's gross weekly earnings remain at approximately two-thirds of men's. This fact alone explains some of the discrepancies in access to mortgages. Building societies prefer to lend to people whose work patterns are likely to be consistent. Women's often more chequered 'careers', interrupted as they tend to be by having to care for others (young and/or, increasingly, old), also disadvantage them in their access to home ownership. In 1987 less than one-third of women with a child under five were in paid work (Labour Force Survey, 1987); nearly 700,000 women were in part-time paid work compared to fewer than 200,000 men; nearly twice as many men as women were gaining extra pay through overtime payments. Thus it is no surprise to find that the households with the lowest incomes are the households headed by women, the lowest-income group being single female pensioners, closely followed by single parents, of whom 89 per cent are women.

The relative number of people living in two-parent/two-children families has continued to decline over the past decade, despite the new Right's rhetoric exhorting a return to traditional values and the Victorian bourgeois nuclear family. The nuclear family household which has so dominated the thinking of policy-makers and politicians is becoming less and less the norm. Taken alongside the increasing numbers of married women in the labour force, the policy discourse is distinctly at odds with the changing times and changing economic necessities. In this context, the promotion of home ownership must be seen as privileging one form of household over another and indirectly as privileging men over women.

Women and rented housing

The opportunities for women to set up independent households, with or without dependants, are being continually eroded. Because of women's lower incomes and employment status, the public rented sector has played a significant role in their housing possibilities. Since the Conservative government has been in power, more than 100,000 council houses every year have been sold into private ownership. The Housing Act 1988 aims to speed up that process with an expected increase in council house sales in 1988–89 of 40 per cent, combined with the push to privatise council estates as a whole. The Housing Act 1988 also aims to make rented housing more expensive and to wind down the statutory obligations of local authorities to provide rented housing for all those in need. The adverse implications of all these changes for women, and for independent women in particular, are obviously enormous.

Single women, both with and without dependants, are particularly reliant on housing allocated on the basis of need rather than according to ability to pay. However, the options available are increasingly those of the private rented sector, despite the fact that this sector represents only one-tenth of the total stock. Research in other countries has shown that estate agents are reluctant to rent to single parents and that women in general meet with considerable discrimination in this tenure (see Watson, 1986). Although a few housing associations do cater specifically for particular women's needs, such commitments are becoming increasingly rare. The commercialisation of housing associations is likely to discourage such commitments in future. There are no specific obligations on the Housing Corporation to promote equal opportunities for women as housing association clients. With regard to the commercial private rented sector, the GLC survey in 1983–84 found that 'women are over-represented in the sector, accounting for an estimated 38 per cent of all heads of households privately renting in London. This implies a total of 125,000 households' (GLC, 1986, p. 19). The survey drew particular attention to the large proportions of non-pensioner women living in the informal rented sector outside the Rent Acts with little security of tenure; they are 'more than twice as likely as the average [private tenant] to be subject to landlord harassment and to have to share basic amenities' (GLC, 1986, pp. 19–20). The government would argue perhaps that this situation may be remedied by the new legislation as market rents protected by housing benefit bring greater landlord investment and a better service for tenants, who also now have greater legal sanctions against landlord harassment. In reality the situation seems likely to worsen; the unregulated private rented sector will continue, tenants' awareness and their ability legally to challenge abusing landlords will not improve much and market rents will exacerbate economic pressures as they did in the wake of similar legislation in 1957.

Overall, therefore, there seems little joy for most independent or potentially independent women in the new policies (see also Morris, 1987; London Housing Unit, 1988, section 3). However, women are in the forefront of the opposition to the government's policies. As in previous struggles, activists in the tenants' movement are predominantly women. The opposition to Housing Action Trusts in Sunderland and Lambeth, for example, has been led by women. The successful defence of subsidised and secure rented housing is vital to women's interests.

Women and homelessness

In the four years from 1984 the number of officially homeless households accepted as being in priority need of assistance in England and Wales increased from 80,000 to 107,000. Women-headed households represent the majority of this group. Similarly the number of homeless households in bed and breakfast, hostels and women's refuges has doubled in the same period to 15,000, of whom more than a third are single-parent families. Austerberry *et al.* (1984) have shown that much of women's homelessness is concealed and that many women never approach local authorities for accommodation, or are not accepted as homeless. The number of households in this latter category was 62,000 in 1987, which is only the tip of a large iceberg. Given that one-fifth of all official homelessness is a result of the breakdown of a relationship with a partner, women going through separation are particularly disadvantaged. Yet the government is clearly hoping that the new legislation will step up the pressure on such women in order to discourage single parenthood and the alleged strategy of having children in order to get a council or housing association tenancy.

In 1988–9 the government reviewed the Homeless Persons Act, which for many is the first rung on the ladder to a decent council home via 'hard-to-let' property. Nicholas Ridley, as Secretary of State for the Environment, said in a speech in 1988 that people with 'roofs over their heads' did not deserve to be counted as homeless and that homeless people allocated council tenancies are 'jumping the queue' (Shaps, 1988, p. 14). With the implied abolition of the local authority housing queue under the 1988 Act, restrictive amendment or abolition of the Homeless Persons Act seemed imminent. Conservative Central Office gave their MPs a

'Homelessness Briefing Pack', which claimed that 'the incidence of homelessness is not the same as a national housing shortage', going on to explain that

> the government's approach is designed to divide homelessness into a number of discrete issues, with a reasonable tale to tell on each, and to avoid treating it in general terms as a large amorphous issue which could only be approached by the injection of unrealistically large amounts of public money.

> (Travis, 1988, p. 6)

The suggestion was that a major cause of homelessness is young people having children in order to secure a council tenancy. This was made explicit by Margaret Thatcher in an interview with *The Times* in which she said she was anxious that ministers act on the growing problem of 'the young single girls who deliberately become pregnant in order to jump a housing queue and get welfare payments' (Travis, 1988, p. 6). Clearly government support for the family does not extend to helping house young parents. The ground was being prepared for the extension of the concept of intentional homelessness to those affected by marital/relationship breakdown, people with illegitimate children, young people living with parents, single parents and so on. Had this succeeded, the vague statutory requirements of local authorities to provide housing for those in need, particularly women, would have been abolished. In the end the government backed off from reforming the Homeless Persons Act, as the high tide of Thatcherism receded.

Housing and older women

Older women are often overlooked when women's housing needs are discussed because of high levels of home ownership amongst this group; approximately 40 per cent of widows own their dwellings outright. Female pensioners will obviously be adversely affected by the privatisation of public housing, since their incomes as a group are the lowest of all households, but not all women home-owners are adequately housed either. Many older, women home-owners cannot afford repairs and are unsure how to manage their finances. They are also fearful of being ripped off and/or harassed by men working on the house. As a consequence, many women live with small repair jobs not done, rather than take the risk of having a man they do not know in their home (see Coleman and Watson, 1987).

Women constitute the majority of older people as a result of their greater longevity. In 1986 there were four times as many widows as widowers aged 65 and over, and in 1987 there were twice as many women over 75 as men. Another important issue, therefore, is the possible need for care in old age.

The Thatcher governments' policies towards elderly people have had several effects. There has been a failure to provide a sufficient stock of sheltered housing, which has increased the likelihood of institutionalisation and dependency. There has also been a deliberate shift towards the privatisation of care, which is unaffordable for many low-income people, combined with a rhetorical emphasis on the desirability of care in the community. Given the lack of funds with which to make this a reality, this shift in effect means that, increasingly, responsibility for the care of older women falls on their daughters or daughters-in-law with all the stresses and strains entailed. Women with disabilities are also disproportionately hit by the decline in public-sector provision, since they are over-represented in this tenure (see Morris, 1988). Moreover, the rights of women who are caring for an older or disabled person to succeed to a private tenancy have been curtailed by the Housing Act 1988.

Housing for divorced and separated women

It is clear from the preceding discussion that women as independent householders are severely disadvantaged. This relates to the second strand of the argument, which concerns the ways in which women's dependence on men, their powerlessness and their entrapment in the domestic world, are both created and reproduced by the housing system in Britain. This becomes particularly clear in cases of separation and divorce, and where there is domestic violence. In 1987 a total of 165,000 decrees for divorce were made absolute, which was double the number in 1971 when the Divorce Reform Act 1969 came into force in England and Wales. Research in the USA by Weitzman (1981), in Australia by Watson (1985) and in New Zealand by Davey (1985) has found that women's economic and housing position tends to decline on divorce, whereas their partners' situation tends to either remain the same or improve. In Britain, whereas a third of divorced or separated men continue to live in a house with a mortgage, only a quarter of such women do. These figures challenge the myth that it is women who tend to end up with the marital home on divorce. Instead, divorced and separated women in the past have been far more reliant on local authority rented accommodation; half of these households lived in this sector in 1984. A study of the housing consequences of marriage breakup in Glasgow (see also Brailey, 1984) concluded that

> even so the freedom and security associated with a home of one's own can make it all worthwhile. For many women the benefits of independence justified all the trials and tribulations of both separation and the process leading to it.
>
> (Brailey, 1986, p. 69)

For women who are older, who have been out of the labour force for some years raising children, or who are in part-time or intermittent paid work, the likelihood of getting finance, of being able to afford the repayments or of being able to afford to buy out their ex-partner is small. In other countries (see, for example, Watson, 1988), state housing authorities have devised schemes to enable women to remain in the marital home after divorce, thereby ensuring less disruption to themselves and their children. These include shared equity arrangements and restructuring mortgages using capital-indexed loans to overcome the high real levels of repayment in the early years of the loan by replacing constant nominal repayments with constant real repayments. For many single mothers, current income is often an inadequate guide as to future income and ability to repay, since their low-income state may well be temporary relative to a long loan period. Pressure for such policies will increase during the 1990s in Britain.

Where women have to leave a relationship because of domestic violence, the housing problems they face can be even starker, particularly if they are forced to leave in a hurry. There has been a substantial increase in the number of refuges since the early 1970s, but there remains a huge shortage of places. Binney *et al.* (1981) found that accommodation problems were the most powerful constraint preventing women leaving violent men. This was particularly the case for the high proportion who had no independent earnings. Once in a refuge the major problem is finding alternative, permanent and affordable accommodation. Given women's exclusion from private housing options, local authority housing departments' responses are extremely important. Local authority policies and practices towards women at risk of domestic violence seem to vary widely. Under the Homeless Persons Act local authorities are obliged to house anyone they deem to be 'vulnerable'. Women without children are usually excluded from this category. Where a woman is classed as 'vulnerable', some condition is usually placed on her before accommodation is provided, for example proof of violence from a GP or initiating divorce proceedings (see Maguire, 1988). Inevitably, the acceptance of a woman as at risk means the housing department bears the cost of bed and breakfast, which is unpopular. Maguire (1988) suggests that part of the problem lies in housing officers' perceptions of beaten women as subordinate, inferior, blameworthy and so on. This problem should be addressed by training programmes in order to dispel such myths amongst housing officers. Clearly also there is a need for more refuges beyond the present number of 230 which has remained roughly static throughout the 1980s, as some have closed due to withdrawal of grant aid and others have opened. Mansfield and Ashfield Women's Aid (1988), for example, have given a vivid account of their recent two-year-long struggle to establish a local refuge. In October 1989, board and lodgings payments to refuge residents will be replaced by Income Support and Housing Benefit, which will lead to a severe loss of income to both residents and refuges. Women's Aid is presently challenging these proposals, which if they are implemented could lead to widespread refuge closures (see Brotchie, 1989). There is also a continuing power struggle for autonomy from local authority control being waged by women in the refuge movement. Mama suggests that

> what is being witnessed is a state encroachment into refuge management, even though it may be through supportive and well-intentioned liaising officers. This encroachment is most evident in refuges for Black women, and the race implications of this deserve particular attention. . . . Local Authority controlled refuges would represent an attack on the autonomy which has been the force behind the refuge movement.
>
> (Mama, 1989, pp. 39–40)

Finally, there is also a need for increased public investment in rented housing for such women in particular, so that returning to a violent man or remaining in an overcrowded refuge for months on end are not the only options. The 1990s will see renewed struggles to further these aims.

Black women and housing policies

It must be apparent from the above that black women suffer a double disadvantage in their housing as in their other welfare needs, though this is, as yet, impossible to quantify adequately. There can be no question that

> black women bear the brunt of poor housing and pay a heavy price in terms of poor physical and mental health, racial and sexual harassment and attacks, and find themselves cut off from the social life, not only of the white community, but of other black communities as well.
>
> (Khanum and Arboine, 1988, p. 1)

Amin, for example, suggests that 'black women experience the brunt of the violence that is directed against the black household' in the form of racial harassment (Amin, 1987, p. 12). Leicester Racial Attacks Monitoring Project also suggest that 'over the last year a clear pattern of racist attacks and harassment against black women has emerged' (LRAMP, 1988, p. 3). LRAMP (1988) give details of eight recent cases in Leicester of attacks against single black women.

In relation to domestic violence against women in the home, Mama describes how in recent years black women have 'organised separate refuges in response to the problem of ethnocentrism and racism within the refuge movement' (Mama, 1989, p. 42), thus developing ways in which black women 'creatively apply themselves to addressing our own needs and to help ourselves when empowered to do so'. Yet, as she argues,

> racism in all the institutions that Black women have to deal with means that they do require extra support and advocacy from refuge workers, not least because they currently tend to be in the refuge for longer periods of time awaiting rehousing.

(Mama, 1989, p. 43)

Over the 1980s in Britain a black women's housing movement has become increasingly vocal and organised. For example, the report of the second Black Women and Housing Conference (see Lambeth, 1988) offers comprehensive, radical policy proposals for the local and national housing policy agenda. Black women have specific and pressing housing needs to which local authorities, housing associations and other housing agencies should address themselves in the 1990s.

A visionary approach

Finally, there is a need for a more visionary approach to housing policy if gender issues are going to have any real meaning. It has been argued that the form of housing in Britain reinforces the dominance of the patriarchal nuclear family household to the exclusion of others. This is as much an issue about the design and form of housing and the urban environment as it is about access and allocation (see Watson, 1988). Feminist architects and planners (see Matrix, 1984) have attempted to challenge the dominant discourses and practices within their professions, but they have had limited power to do so and therefore limited effect. In thinking about housing policies for the 1990s, we should challenge the very basis of the way that housing is provided and produced, and look back to the communitarian and utopian socialist thinking of the early to mid-nineteenth century (see Hayden, 1981), which placed the built environment so centrally in its visions for a more egalitarian and non-sexist world.

CONCLUSION

The housing situations of black people and of women in Britain have not been focuses of government policy in the past and are unlikely to be in the near future. Occasionally back-bench MPs on select committees and government reports have expressed concern about racial inequalities in housing, re-housing battered women and the housing of single mothers. Yet explicit Labour or Conservative government policies or initiatives on such issues have been entirely absent. The key housing policy documents of the past two decades – the Labour government's Green Paper (DoE, 1977) and the Conservative government's White Paper (White Paper, 1987) – have virtually nothing to say about the issues raised in this chapter. It will be a hard struggle to put them onto the national political agenda in the 1990s. At the local level, the now long-established struggles on such issues will continue.

NOTES

1 In this chapter the phrase 'black people' is used to describe people of Afro-Caribbean, African and Asian origin.
2 In this chapter the Housing Act 1985, Part III, formerly the Housing (Homeless Persons) Act, 1977, is referred to simply as the Homeless Persons Act.
3 For an expanded version of this review of institutional racism and local authority housing, see Ginsburg (1989a).
4 This is based on a wider discussion of the Housing Act 1988 in Ginsburg (1989b).
5 A more extensive version of this discussion of racial harassment and housing policy is offered in Ginsburg (1989c).

REFERENCES

Amin, K. (1987) 'Black Women and Racist Attacks', *Foundation*, no. 2, pp. 12–15.
Austerberry, H., Schott, K. and Watson, S. (1984) *Homeless in London 1971–1981*, London: London School of Economics, STICERD.
Austerberry, H. and Watson, S. (1981) 'A Woman's Place: A Feminist Approach to Housing in Britain', *Feminist Review*, no. 8, pp. 49–62.
Binney, V., Harkell, G. and Nixon, J. (1981) *Leaving Violent Men*, London: Women's Aid Federation of England.
Bonnerjea, L. and Lawton, J. (1987) *Homelessness in Brent*, London: Policy Studies Institute (PSI).
——(1988) *No Racial Harassment This Week*, London: PSI.
Brailey, M. (1984) 'A Woman's Place is in the Home', *Roof*, September–October, pp. 19–21.
——(1986) 'Splitting Up – and Finding Somewhere to Live', *Critical Social Policy*, no. 17, Autumn, pp. 61–9.
Brotchie, J. (1989) 'Refuge Without Assurance', *New Statesman and Society*, 7 April, p. 31.
Brown, C. (1984) *Black and White Britain: The Third PSI Survey*, London: Heinemann.
Coleman, L. and Watson, S. (1987) *Women Over Sixty*, Canberra: Australian Institute of Urban Studies.
Commission for Racial Equality (CRE) (1984) *Race and Council Housing in Hackney: Report of a Formal Investigation*, London: CRE.
——(1987) *Living in Terror: A Report on Racial Violence and Harassment in Housing*, London: CRE.
——(1988a) *Homelessness and Discrimination*, London: CRE.
——(1988b) *Racial Discrimination in a London Estate Agency: Report of a Formal Investigation into Richard Barclay and Co.*, London: CRE.

——(1988c) *Parliamentary Report, 1988: The Housing Bill*, London: CRE (mimeo).

——(1989) *Racial Discrimination in Liverpool City Council: Report of a Formal Investigation into the Housing Department*, London: CRE.

Dame Colet House (no date) *Tenants Tackle Racism*, London: Dame Colet House.

Davey, J. (1985) *Marriage Breakdown and Its Effect on Housing*, Wellington: National Housing Commission Research Paper 85/2.

Department of Employment (1987a) *New Earnings Survey*, London: HMSO.

——(1987b) *Labour Force Survey*, London: HMSO.

Department of the Environment (DoE) (1977) *Housing Policy: A Consultative Document*, London: HMSO.

——(1986) *General Household Survey 1984*, London: HMSO.

——(1988) *English House Condition Survey 1986*, London: HMSO.

Equal Opportunities Commission (EOC) (1978) *It's Not Your Business: It's How Society Works: The Experience of Married Applicants for Joint Mortgages*, Manchester: EOC.

Forbes, D. (1988) *Action on Racial Harassment: Legal Remedies and Local Authorities*, London: Legal Action Group.

Forman, C. (1988) 'Cooperative Housing – One Way Forward?', *Foundation*, no. 4, June, pp. 14–17.

Ginsburg, N. (1989a) 'Institutional Racism and Local Authority Housing', *Critical Social Policy*, no. 24, Winter 1988–9, pp. 4–19.

——(1989b) 'The Housing Act, 1988 and Its Policy Context: A Critical Commentary', *Critical Social Policy*, no. 25, Summer, pp. 56–81.

——(1989c) 'Racial Harassment Policy and Practice: The Denial of Citizenship', *Critical Social Policy*, no. 26, Autumn, pp. 66–81.

Grant, C. (1988), 'Raising the Stakes', *Roof*, November–December, pp. 30–3.

Greater London Council (GLC) (1986) *Private Tenants in London*, London: GLC.

Hayden, D. (1981) *The Grand Domestic Revolution*, Cambridge, MA: MIT Press.

Henderson, J. and Karn, V. (1987) *Race, Class and State Housing*, Aldershot: Gower Press.

Home Affairs Committee (1987) *Bangladeshis in Britain: First Report from the House of Commons Home Affairs Committee 1986–7*, volume 1, London: HMSO.

Home Office (1981) *Racial Attacks: Report of a Home Office Study*, London: HMSO.

Jacobs, S. (1985) 'Race, Empire and the Welfare State', *Critical Social Policy*, no. 13, pp. 6–28.

Karn, V. (1983) 'Race and Housing in Britain: The Role of the Major Institutions', in Glazer, N. and Young, K. (eds) *Ethnic Pluralism and Public Policy*, London: Heinemann.

Karn, V., Kemeny, J. and Williams, P. (1985) *Home Ownership in the Inner City*, Aldershot: Gower Press.

Khanum, N. and Arboine, J. (1988) 'Black Women In Housing', unpublished manuscript.

Lambeth (1988) *Black Women Challenge Oppression and Organise for Change: A Report of the 2nd Black Women and Housing Conference*, London: London Borough of Lambeth.

Lawrence, E. (1987) 'Editorial', *Foundation*, no. 3, December, pp. 3–7.

Leicester Racial Attacks Monitoring Project (LRAMP) (1988) *Progress Report no. 3*, Leicester: LRAMP.

London Against Racism in Housing (1988) *Anti-Racism for the Private Rented Sector*, London: London Against Racism in Housing.

London Housing Forum (1988) *Speaking Out: Report of the London Housing Enquiry*, London: London Housing Forum.

London Housing Unit (1988) *Just Homes? The Equal Opportunity Implications of the Housing Bill*, London: London Housing Unit.

London Research Centre (1989) *The London Housing Survey 1986–87: Full Report of Results*, London: The London Research Centre.

Luthera, M. (1988) 'Race, Community, Housing and the State' in Bhat, A., Carr-Hill, R. and Ohri, D. (eds) *Britain's Black Population: A New Perspective*, Aldershot: Gower Press.

MacEwen, M. (1988) 'Racial Incidents, Council Housing and the Law', *Housing Studies*, vol. 3, no. 1, pp. 34–45.

Maguire, S. (1988) 'Sorry Love: Violence Against Women in the Home and the State Response', *Critical Social Policy*, no. 23, Autumn.

Mama, A. (1989) 'Violence Against Black Women: Gender, Race and State Responses', *Feminist Review*, no. 32, Summer, 30–48.

Mansfield and Ashfield Women's Aid (1988) 'Last Night I Heard the Screaming', *Roof*, July–August, 27–8.

Matrix (1984) *Making Space: Women and the Man-Made Environment*, London: Pluto Press.

Morris, J. (1987) *The Impact on Women of National Housing Policy Since 1979 and Prospects for the Future*, London: Shelter.

——(1988) 'Keeping Women in their Place', *Roof*, July, pp. 20–1.

Phillips, D. (1986) *What Price Equality?: Report on the Allocation of GLC Housing in Tower Hamlets*, London: Greater London Council.

——(1987) 'Searching for a Decent Home: Ethnic Minority Progress in the Post-War Housing Market', *New Community*, vol. XIV, no. 1/2, pp. 105–17.

Racial Attacks Group (1989) *The Response to Racial Attacks and Harassment: Guidance for the Statutory Agencies*, London: Home Office.

Shaps, M. (1988) 'Rubbishing the Act', *Roof*, November–December, pp. 14–16.

Simpson, A. (1981) *Stacking the Decks: A Study of Race, Inequality and Council Housing in Nottingham*, Nottingham: Nottingham Community Relations Council.

Travis, A. (1988) 'Time for Cathy to Come Home', *Guardian*, 11 November.

Ward, R. (1982) 'Race, Housing and Wealth', *New Community*, vol X, no. 1, pp. 3–15.

Watson, S. (1985) *Housing After Divorce*, Matrimonial Property Law Research Paper no. 2., Sydney: Australian Law Reform Commission.

——(1986) 'Women and Housing or Feminist Housing Analysis?', *Housing Studies*, vol. 1, no. 1, pp. 1–10.

——(1988) *Accommodating Inequality: Gender and Housing*, Sydney: Allen and Unwin.

Watson, S. with Austerberry, H. (1986) *Housing and Homelessness: A Feminist Perspective,* London: Routledge.

Weitzman, L.J. (1981) 'The Economics of Divorce: Social and Economic Consequences of Property, Alimony and Child Support Awards', *University of California at Los Angeles Law Review*.

White Paper (1987) *Housing: The Government's Proposals*, Cmnd. 214, London: HMSO.

ACKNOWLEDGEMENT

While remaining entirely responsible for its content, we would like to acknowledge the advice and assistance of Jheni Arboine in writing this chapter.

8   Council tenants

Sovereign consumers or pawns in the game?

Johnston Birchall

From the point of view of the tenant, which we shall be taking in this paper, the structure of public-sector housing has always been deeply flawed because it has always, as a matter of course, excluded the interest of the users. Council tenants are at least as interested as councillors and housing workers in the use value of their homes, and they make a continual financial commitment in rent. It might be expected that they should be largely in control of their own destiny, since 'most of the costs and benefits of housing accrue to the user and not to the community at large' (Clapham *et al.*, 1990, p. 244). Yet they have traditionally had few rights and have had to rely on channels of communication and representation designed for citizens in general, rather than for consumers in particular. Housing workers have had their interest protected through trade unions and, more recently, through the power that comes through professional knowledge and training, while councillors have seen their interest as being to represent citizens in general rather than just those who happen to be council tenants, and when there has been a conflict between the two, they have generally chosen to represent the wider interest. How did this 'flaw' in the structure come about?

In the early days of municipal socialism, Sydney Webb argued that multi-purpose local authorities should be created, which would provide a massive, organised and planned response to housing needs. Yet his view of the interests involved in the housing process was simplistic; there was only one interest to be taken into account, that of 'citizen-consumers' (Webb, 1891). Now, as some recent commentators have argued, 'the difference between consumers and citizens is all important' (Klein, 1984, p. 20), and '[I]t is necessary to strengthen the position of both in the public services without pretending that they are always necessarily the same' (Deakin and Wright, 1990, p. 12). Webb had his critics at the time, notably G.D.H. Cole. But Cole was concerned to stake the claim of the worker interest, and he also ignored the problem of how consumers in particular would be represented in the decision-making process (Cole, 1980). It was left to Beatrice Webb to start from the point of view of the consumer and offer a balanced view of the representation of citizen, consumer and producer interests. She made a fundamental distinction between voluntary and compulsory consumers' associations, arguing that compulsory association is only justified under the following circumstances (Webb and Webb, 1930):

1 When there is no identifiable constituency of consumers;
2 Where a service is paid for by all, but only used by a minority at any one time;
3 Where the services provided are inter-dependent and cannot be separated;
4 Where compulsory taxation of non-users is required;
5 Where compulsory regulation is needed of anti-social conduct; and
6 Where there is a natural monopoly over a resource.

Clearly, council housing is unlike street lighting or environmental health, because there is an identifiable constituency of housing consumers. It is unlike education or the health service, because the service is not paid for by all but by a minority who use it continuously. This service not only can but, in the views of critics, *should* be separated from other services, in order to give a clearer housing focus (Audit Commission, 1986). Compulsory taxation of non-users is now only required at the national level (since the ring-fencing of Housing Revenue Accounts), and increasingly in the form of Income Support through Housing Benefit rather than through housing subsidy (see Malpass, 1990). Unlike fly-tippers or sellers of faulty goods, council tenants are not an anti-social group who must be compulsorily regulated. Nor is there a natural monopoly, as in gas, water or electricity; there are and should be other providers. It is clear that given the continued shortage of affordable housing, the compulsory co-operation of citizens might be needed to make land, loan finance and public subsidies available; decisions over public resource allocations are unavoidably political. But, on all the above counts, the voluntary co-operation of dwellers ought to be the method by which social housing is provided. The citizen would, through councillors, have a strategic and local planning role which, where necessary, initiates the formation of consumer organisations, such as community housing associations and co-ops, and regulates them to ensure value for money, equal access for minority groups, priority for those in greatest need, and so on. But such organisations would steer clear of the actual provision of housing (see Clapham, 1989).

It is Sydney Webb's rather than Beatrice Webb's model which has been followed, and the consequences for council tenants have been dire.

THE RIGHTS OF TENANTS

An interest implies a right in something. What sort of rights have council tenants had during the 1980s, and what can they expect from the last decade of the twentieth century? A study undertaken at the beginning of the 1980s (to be referred to as the City University study) said:

> Traditionally council tenants have had few legal rights, whereas by comparison, their local authority landlords have had the power to impose a wide range of restrictions on tenants' freedoms to occupy and use their homes.
>
> (Kay et al., 1986, p.vii)

As we shall see, tenants have gained and lost a few rights since then, but only the 1990s will tell whether their position is more secure at the end of the century than it was, under Sydney Webb's benevolent gaze, at the beginning. The interests of tenants can be seen in terms of fundamental rights in use, which have to be enshrined preferably in law but at least in good practice which is backed up by effective sanctions against landlords who violate the rights. Then they have to be communicated to the tenant in a form which is understandable, and be capable of being enforced, with tenants having enough resources to be able to challenge the interests of the landlord if necessary (Kay *et al.*, 1986).

There are *individual* rights applicable to households, and *collective* rights applicable to groups of tenants, ranging from estate level to area level, then to the tenants of a particular landlord and even to national-level negotiations. Then there are rights which are primarily concerned with the relationship of landlord and tenant and what can be done about it if it goes wrong, and rights concerned with different aspects of the housing process. The former include the rights to security of tenure; to a well-defined and equal landlord–tenant relationship as summed up in a contractual tenancy agreement; to information and consultation over housing policies and practices; to negotiation or bargaining over important decisions; to change the relationship by developing various types of joint or self-management; and finally to opt out of it altogether by changing landlords or transferring ownership to a tenant-controlled body. The latter include rights over access to the tenancy and mobility within it, rights concerning rents, management and maintenance, and rights over the design and modernisation of the home. While both types of rights are equally important, in this paper only the former – the rights concerning the relationship between landlord and tenant – will be considered in detail.

Security of tenure

Probably the most fundamental right a tenant can have is security of tenure. Without this there are no other rights, because the rights that remain become conditional on the landlord's goodwill. Yet, until 1980 council tenants were denied a basic right which had been given to most private tenants and which was inherent in the legal status of an owner-occupier. In addition, they were not guaranteed the right as members of a household to succeed to the home; most landlords granted it, but only at their discretion. Owner-occupiers took it for granted that they could leave the home to anyone they chose, while private tenants, at least until the 1988 Housing Act, had the right of two successions for relatives, provided they had been living in the home for at least six months. Yet prior to 1980 all that a local authority had to do to gain possession of a council house was to issue a notice to quit which ended the tenancy 28 days later, then go to court for legal authority to repossess.

The decade started badly for council tenants; the response of landlords to the proposed 'Tenants Charter' of the 1980 Housing Act was almost completely negative. Both the Association of Metropolitan Authorities (AMA) and the Association of District Councils (ADC) were against security of tenure. First, they used the 'democratic accountability' argument: that security was not necessary because local democracy was 'sufficient safeguard against any abuses' (Laffin, 1986, p.194). Yet during the 1970s there had been several notorious cases where tenant activists had been threatened with eviction by landlords prepared to push their powers to the limit (Ward, 1974). Second, the associations used the 'responsibility to citizens' argument: that security would undermine their freedom to manage their property not on behalf of the individual tenant but on behalf of consumers and citizens in general: 'The public sector landlord owes a duty not only to the tenant in question but to all his other tenants and to the general public, to see that the best use is made of the housing stock' (quoted in Laffin, 1986, p. 202).

The security granted by the 1980 Act was a compromise with the landlord. It forced the housing manager for the first time to prove to a court that a breach of tenancy (such as rent arrears or nuisance) had occurred, but there was still a mandatory ground for possession when modernisation was planned, or when the dwelling became overcrowded. But the tenant was now secure until proved otherwise. The Act also gave one right of succession, to the spouse or another member who has resided in the home for at least twelve months before the death of the tenant. The landlord was given the right to transfer non-spouse successors between six and twelve months from the start of the succession, if they were under-occupying.

Difficulties remained, however, for co-habiters and for those who could not prove they had lived in the home continuously for the twelve months, and if there were more than one claim to succession.

Clearly, landlords were still fully in control of their 'property'. Yet, even so, they were reluctant to ensure that the tenants knew their rights; local authority tenancy agreements often still implied that there was no security, 30 per cent still made no mention of the grounds for repossession at all, and only a third mentioned the four-week-notice process. More seriously still, 'The majority of authorities failed to explain to their tenants that they could now defend themselves in court' (Kay *et al.*, 1986, p. 80, figures from a survey of England and Wales). Less than half (47 per cent) of authorities provided an accurate description of the statutory right to succession.

With hindsight, the Conservative government might well regret the granting of even such a basic right as security of tenure. During the second half of the decade, ministers set out to break up council monopolies, and security got in the way. Partnership schemes with private developers began to be held up, and a new clause had to be added to the 1986 Housing and Planning Act, allowing councils who wish to sell estates to evict tenants, offering alternative accommodation. Then they made the fatal error of trying to persuade council tenants to opt for other landlords at the same time as making these alternatives less secure.

Tenants have rejected Housing Action Trusts, and some voluntary or Tenants' Choice transfers to housing associations, partly because they promised less secure 'assured tenancies'. Like the local government associations in 1980, central government in the late 1980s has under-estimated tenants' need above all for security, and has in consequence experienced a whole series of setbacks to its housing strategy.

A fair and equal landlord-tenant relationship

On entering into a contract, both parties usually try to make sure that they know what they are getting into and that the relationship will be a fair and mutual one. However, most prospective tenants are not in a strong position to argue over the terms of their tenancy, or even to ask for clarification of details; their overwhelming need is to be housed. The result is that 'the provider of housing therefore has a great advantage which he can, if he wishes, exploit' (NCC, 1976, p.7). During the 1980s, a consensus emerged that tenants should have a clear, simply worded tenancy agreement which sets out the contractual relationship between landlord and tenant, supplemented by further useful information provided in a handbook. The rights and duties of each party should be set out in a fair and balanced way, and there should not be petty restrictions on the tenant, nor an attempt by the landlord to avoid legal responsibilities to repair the dwelling. There should be a clear procedure for dealing with tenants' complaints, arbitration services to resolve disputes, and compensation when damage or delay has been caused by the landlord (compare Housing Corporation, 1989).

Before the 1980s began, there was a well-documented critique of the unequal nature of tenancy agreements (Fox, 1973; Housing Services Advisory Group, 1978), coupled with condemnation of the way in which housing visitors were being used as spies to harass tenants into complying (Ward, 1974). A National Consumer Council (NCC) study gained much media attention by providing colourful examples of incredibly petty and insulting clauses in tenancy agreements. It found that agreements were generally one-sided, imposing obligations on tenants but not on landlords, and that councils were wrongfully excluding their liabilities in law; 87 per cent did not mention the responsibility for repairs. The pervasive mood of many of the agreements was paternalistic, 'more suited to the last century than this', and they were punitive; 64 per cent said they would terminate the tenancy if the agreement were breached, and some implied that tenants would immediately be ejected from the home without notice. Lastly, the agreements were incomprehensible, laid out in legalistic language which tenants could not be expected to understand (NCC, 1976, quote from p.12).

It seems incredible that despite this accumulation of evidence, the right to a written tenancy agreement was not given to English and Welsh tenants in the 'Tenants Charter' of the 1980 Housing Act (it was, however, given to Scottish tenants, who can now challenge their agreements in court, though no-one has yet done so). What was given to tenants in England and Wales was the right to a minimum amount of information on terms of tenancy, which did not include information on other landlord obligations in law, nor on the remedies available to tenants if the landlord breaches the obligations it does have. All that the landlord must do is inform the tenant of the new 'Tenants Charter' rights under the 1980 Act and provide a summary of the repairing obligations under the 1961 Act.

The local authorities' response was slow and grudging. The City University study found that only 70 per cent of them met the time limits for providing information on terms of tenancy, and only 52 per cent on the issue of explanatory information. Most did remove the more petty restrictions, but many examples of restrictive rules remained, such as those against hanging out washing or putting nails in walls. Yet 42 per cent were found still to be unbalanced, emphasising prohibitions on the tenant, and a significant minority still failed to mention some of the tenants' rights: the right to buy, to consultation, or to succession. Over half qualified their obligations in some way, in some cases illegally! Only 29 per cent gave a full list of the landlord's obligations, and more than half did not mention that common parts in blocks of flats are the landlord's responsibility. Few local authorities had considered what to do if things went wrong: only 4 per cent offered compensation for failure in services, 9 per cent had an appeals procedure, and only 2 per cent an arbitration panel. The conclusion was that 'authorities have not gone out of their way to provide useful information to tenants beyond that legally required' (Kay *et al.*, 1986, p. 67). Nor did the responsible central government body, the Department of the Environment (DoE), go out of its way to provide a model tenancy agreement which might set out good practice in a clear and uncompromising way.

Changes in practice since then have been led by the housing association movement; in 1987 the National Federation of Housing Associations (NFHA) provided a model tenancy agreement, setting out a 'clear, legally correct, comprehensive and workable basis for the legal relationship between the association and its tenants' (NFHA, 1987, p. 42), and in 1989 the Housing Corporation published its expectations that associations will comply with a stiff code of practice. Such clear performance expectations for associations call into question the continuing lack of monitoring powers over local authority practices. It may be that, under the impetus of Tenants' Choice and the process of obtaining assent to 'voluntary transfer' of housing stocks, many local authority agreements are also being reappraised. Individual local authorities have begun to experiment with innovative procedures: Southwark Borough Council has instituted independent local arbitration under the Tribunal Act 1974, while Halton Borough Council has an appeals panel which includes tenant representatives (Platt *et al.*, 1990). But information about the global situation is not available; part of the problem is that, because of the lack of central monitoring and the lack of a survey since the all-important 1988 Housing Act, there is no hard information available on which we can judge current performance.

What, then, should be the role of tenants' organisations in improving and safeguarding this right? In 1980, only 22 per cent of councils involved tenants' representatives in the drafting, and 33 per cent in commenting on the final draft, of the tenancy agreement. Experience from the ones that did showed, not surprisingly, that the more tenants are involved the more their interests are protected. The conclusion from researchers at the time was that '[p]aternalism may have been dented, but it has not been eradicated or even reduced in certain places' (Kay *et al.*, 1986, p.70). Yet, if tenants are not involved, it can hardly be called a tenancy *agreement*.

Information

The right to give and receive information is a basic part of any relationship, including that between landlord and tenant; without it, arguably there is no relationship. There is certainly no way of ensuring that genuine consultation takes place. As one study puts it: 'Consultation is likely to be unproductive and frustrating for all concerned if tenants . . . are not given sufficient information to make a sensible contribution' (NFHA, 1987, p. 26). What sort of information have tenants been able to claim as of right? The agenda and most reports and background papers for housing committees must be made available to tenants before committee and sub-committee meetings, and minutes must be published after a meeting (1972 Local Government Act). But in practice, one study found that often when tenants want to see the papers relating to their estate or home, they are told they have no rights to see them; this applies to 90 per cent of housing department paperwork, because only those matters considered by councillors become public documents (Bartram, 1988). The Housing Corporation now expects associations to consider ways of giving tenants information about the association in the form of a newsletter or annual report, saying who the committee members are, how to become a member, who is responsible for managing the home, and so on. Guidelines state that they should make available to tenants, on request, copies of non-confidential minutes, agendas and reports, provide clear notification of forthcoming meetings, give tenants the right to attend as observers on boards or sub-committees, and keep a register in the office available to tenants setting out the names and addresses of committee members (Housing Corporation, 1989).

Local authorities are not under clear instruction, and so reformers have had, in this as in so many cases, to rely on exhortation and example. For instance, the Community Rights Project cites Leeds City Council's policy that any tenant has the right to see any piece of paper in the Housing Department unless it is genuinely confidential; in at least nine authorities there is now a presumption in favour of openness (Bartram, 1988). More generally, a major study of tenant participation in Britain (to be referred to as the Glasgow participation study) has recently found that the form in which information is given tends to be impersonal and targeted at individual households rather than tenants' groups or particular estates. By 1987, although most authorities were producing written material, it was mainly in the form of letters, handbooks or leaflets (Cairncross *et al.*, 1989). The quality of the information was often poor. A Scottish survey agreed, and also found that more than half the material studied contained clumsy English and legal or technical language (Goodlad, 1986). A major survey of housing management in England (to be referred to as the Glasgow housing management study) found that, by 1986, only half (51 per cent) of council tenants felt they were kept informed by their landlord, compared with four-fifths (81 per cent) of housing association tenants. Evidence from tenants' discussion groups showed that the information provided was 'often confined to rent increases and was frequently regarded as unintelligible' (Maclennan *et al.*, 1989, p. 93).

All this is about to change. The 1989 Local Government and Housing Act has given council tenants a potentially powerful right to receive annual information of housing performance. The information must be readily understood and cover the broad scope of the landlord function; it must be of the kind that is most relevant to tenants' interests; it must permit a realistic and relative assessment of performance; and it must cover quality as well as quantity of service. In due time it should include estate-based data, 'to give tenants an appreciation of performance in their own local environment' (DoE, 1990). Included will be information on the context in which the landlord works, on rents and arrears, the costs of and time taken on repairs (with assessment of satisfaction), allocations and voids, Housing Benefit claims, and statutory homelessness. The information will estimate average management costs and show the costs of different services. It will describe the procedures for dealing with complaints and arrangements for consulting tenants, and it will promote the participation of tenants in the management of dwellings.

This all sounds quite revolutionary in relation to the paltry amount of information currently available to tenants. But will it be used? In a study of similar performance indicators in other service areas, Day and Klein state, 'There is . . . little evidence that either authority members or the public at large respond to or use such information' (1987, p. 243). In housing, however, there is the advantage that consumers can be identified who are not just 'members of the public' and who have a direct interest in using such information in their negotiations with the landlord. Clearly, its use is only going to be as good as the level of effective organisation which tenants have reached. Used effectively, it has the potential to transform the power relationship, but also to sharpen the division of interests between landlord and tenant; as one tenant representative on a housing association committee said recently, 'Now that we're getting to know more, they don't like it much' (NFHA, 1987, p.11).

Rights to receive information from the landlord also begin to show up the need for independent sources of information, particularly if the government is trying to persuade tenants to accept a Housing Action Trust, or the council is trying to persuade tenants to accept a voluntary transfer to a new landlord, or hoping for their loyalty in relation to Tenants' Choice. Recently there has been a growth in such information provision, both from officially sponsored organisations such as the Tenant Participation Advisory Services, and from tenant-controlled organisations such as the

Tenants' Resource and Information Service. This was launched in September 1990 as a specialist agency, set up and controlled by tenants, who say 'we want to be able to provide these resources and control the services ourselves' in order to provide independent information on the choices available (*London Housing News*, 1990). A similar organisation, the Tenants' Information Service, was set up for Scotland in mid-1989; its priority is to be an information centre providing for the tenants' movement's information and training needs. The fact that a library, newscuttings file and database are part of the organisation's priorities shows the extent to which tenants, working in a changing environment, see the need for information before anything else.

Consultation

Consultation can be defined as 'the process of asking tenants for their views in order to consider them before a decision is made' (Cairncross *et al.*, 1989, p. 64). It can be undertaken at the level of individual households or collectively, through tenants' associations and forums. At first sight, the *individual* right to consultation appears an uncontentious one which, when it was demanded, attracted all-party support and then was enacted in the 1980 Tenants' Charter for both local authority and housing association tenants. In fact, it was the subject of much debate. The 1977 Labour Green Paper proposed a right to consultation on changes in rents and other charges, in policies and practices, in maintenance and improvement programmes, and in housing management, provided that tenants are substantially affected. The Conservative government took over the Labour proposals, yet, under pressure from all three local authority associations, limited the right to changes in housing management and improvement programmes, but excluding rent levels. The definition of what changes 'substantially affected' tenants was left to the landlords, as was the method for consultation; landlords were charged simply to 'maintain such arrangements as they consider appropriate' and then publish them (Housing Act 1985, section 105). The NCC was 'dismayed' at the weakness of requirements which gave the landlords so much discretion.

The local authority associations had objected to the right on the grounds that it would be costly and administratively difficult to put into practice, and so it is not surprising that the landlords showed little enthusiasm for their new responsibility; the City University study found that only 50 per cent met the deadline set for publishing their consultation statements, and only 31 per cent more met a second deadline a year later. Only a small minority (14 per cent) had made any strong, formal commitment to consultation; few had told tenants where to write with their comments; even fewer had given a commitment to consider their views. Only 9 per cent said they would inform tenants of the outcome of a consultation exercise and, incredibly, 44 per cent had failed to consult tenants over management changes that had taken place since the 1980 Act came into force (Kay *et al.*, 1986).

By 1986, the Glasgow housing management study had found little improvement. Only a fifth of council tenants believed the landlord always or usually consulted them on important issues, and only half of those who had been consulted believed the landlord had taken heed of their views following consultation. Yet there was a widespread desire for more; 70 per cent of tenants wanted more influence on the landlord's decisions. Discussion groups illuminated how tenants felt: they resented the lack of consultation, and where it did take place it was confined to modernisation or new building proposals. Tenants also wanted formal consultation on repairs, allocations, and rent setting (Maclennan *et al.*, 1989).

What about *collective* consultation? The 1980 Act was criticised at the time for reducing collective rights to individual ones, in a 'shift away from and dilution of the broader proposals that tenants' organisations were making ten years earlier' (Kay *et al.*, 1986, p.15); the argument is that only through collective action can tenants gain the influence they need to make sure their voice is heard. Such action had been gaining ground before 1980, but only very slowly. A Labour government consultation paper had expressed ministerial doubts about 'the real willingness of local authorities to set up voluntary schemes of tenant involvement', and suggested a statutory tenants' committee as a way of forcing the issue. The idea was a radical one, that such committees would be set up in every local authority, from estate-level upwards, and would consider all issues including both management and the more contentious areas of rents, allocations policies and transfers. The landlord would be bound to consider any suggestions and to respond to them. Once again the idea was widely attacked by the local authority associations and professional bodies, and the eventual Conservative Act dropped the idea of the collective right to a statutory consultative committee; as the City University researchers put it, 'local authorities could be pleased that their very active lobbying over a period of years had blunted the threat to their loss of power' (Kay *et al.*, 1986, p.15).

During the 1980s, the rhetoric changed to a widespread support for collective tenant participation; virtually all the studies of housing policy and practice advocated it (NFHA, 1985; Audit Commission, 1986; etc.), and Tenant Participation Advisory Services were gradually set up, first in Scotland in 1980, then in England, Wales and Northern Ireland, with some grant aid from the Conservative government. Local authorities and tenants' groups were able to draw on these for expert guidance in setting up participation schemes. By 1987, the Glasgow participation study found that two-thirds (69 per cent) of landlords were convening irregular discussion meetings, around a third (35 per cent) regular discussion meetings, and 23 per cent advisory committees. There is no doubt that the use of these methods grew substantially throughout the 1980s, but that the more formal and regular the scheme, the fewer the number of councils who were offering it; involvement of tenants on decision-making committees with advisory or observer status was still very restricted, though the proportions had doubled to 8 per cent since the City University study in 1981.

The *level* at which schemes operate has changed significantly, with decentralisation of services and of consultation being a feature of the 1980s. During the 1970s, there had not been much real decentralisation; Richardson had found that, of fifty local authorities which had participation schemes, only six were estate-based (1977). However, ten years later the Glasgow participation study found that around a third (31 per cent) of authorities had some kind of estate-based initiative with tenant involvement (Cairncross *et al.*, 1989). A consensus emerged that decentralisation of housing services was a good thing; the Audit Commission, for instance, recommended that delegation should be as far down the line as possible, preferably to estate level, in order to increase the level of service, tenant involvement and satisfaction (1986). However, Clapham notes that although decentralisation is an essential ingredient in user control, the trend has not been accompanied by genuine devolution of decision-making. Plans to set up local committees to control decentralised offices have failed, because 'the problems of agreeing the appropriate make-up and terms of reference of the committees have proved in most cases to be insurmountable' (Clapham, 1990, p. 67). Councillors have been reluctant to devolve power, and the changes have been imposed from the top down with tenants rarely being consulted about the implementation of decentralised services.

What *range of issues* has been covered by consultation schemes? The City University study found that, of 192 statements on consultation which they examined after the 1980 Act came into effect, most mentioned changes in management and new programmes of work, but only one local authority gave a practical list of all the topics it would allow to be raised. Gravesham District Council expressed a mixture of paternalism and willingness to listen which illustrates the dilemma they felt they were in, wanting to control the discussion and yet not be seen to be unreasonable:

> . . . it is for the Council to decide in any particular case whether

consultation is necessary or appropriate, but the council would not wish knowingly or willingly to deny tenants an opportunity to comment on proposals which are of direct personal concern.

(Kay *et al.*, 1986, p.193)

Ominously, 59 per cent of landlords said they would definitely not consult over rents or service charges, and only four London boroughs positively said they would.

Yet since then tenants have managed gradually to widen the range of issues on the agenda. The Glasgow participation study found that, of those authorities which consulted tenants at all, most (81 per cent) consulted on modernisation or rehabilitation. This might seem a high figure but, as it has always been the most widely used and accepted form of consultation, one that is agreed to be essential if the landlord is to ensure value for money and co-operation in gaining access to homes, it is still quite disappointing. Next comes another issue of vital importance to tenants: estate management; only two-thirds of authorities (63 per cent) consulted on it. Next comes consultation – that is, consultation on the form which consultation ought to take. Surely this ought to be 100 per cent! But no: only around half of councils engaged in it. This means that only half could be sure that the consultation procedures they had set up were what the tenants actually wanted. Then comes repairs policy, the favourite subject of tenants and therefore understandably high on their agenda, but it is only on the agenda of a third (34 per cent) of councils. Consultation on new building is difficult but not impossible, yet less than a third of landlords were doing it. Similarly, other issues important to tenants were put on the agenda by a very small minority of landlords.

There may have been improvements in these figures since the study was done in 1986–87. Yet there does not seem to be any evidence to depart from the judgement made in 1981 by the City University group. They found from their case studies that the process of consultation produced well-informed, articulate tenants

. . . who inevitably became involved in, and had strong views on, the political choices raised by council policies. This type of discussion was sometimes resented by or threatening to certain councillors and officers who considered that such decisions should either be left to the politicians or to professional housing officers.

(Kay *et al.*, 1986, p. 208)

Their conclusion is also an explanation:

. . . it was not always easy to cope with the differences in interest that emerged. Many authorities prefer to avoid such debates and to retain complete control over the decision-making process.

(Kay *et al.*, 1986, p. 208).

The final test of the seriousness with which landlords have treated tenant consultation is, of course, the amount of *resources they are prepared to make available* to tenant groups. Where that basic building block of a tenants' movement, the tenants' association, has to be stimulated into being, it makes sense to begin by providing tenant liaison workers. During the early 1980s, the City University study found that this was the most common method of support; 30 per cent of the local authorities in England who responded to their survey employed workers, while 5 per cent gave grants to tenants' associations to employ their own. In addition, 25 per cent supported associations with cheap meeting-places, the use of a duplicator or free printing, while 11 per cent gave grants, including start-up grants to new groups. The majority of councils, however, did not give any support at all to tenants. Only in one authority did tenants have access to a truly independent source of funding: Sheffield had introduced a rent levy by which, for the first time, tenants were able to fund their own activities on a firm basis, without relying just on the goodwill of the landlord (Kay *et al.*, 1986). In Scotland, a similar situation was found, with an active minority of landlords (28 per cent) supporting the development of tenant groups (Community Development Housing Group, 1986).

By 1986–87, the Glasgow participation study had found little change. Because it included Welsh and Scottish authorities along with the English, the figures actually went down; only 18 per cent provided specialist staff, 15 per cent provided premises, 8 per cent starter grants and 8 per cent other grants. Again, the majority of councils (76 per cent) provided no support at all. The level of support was closely correlated with the size of the housing stock and the number of tenants' associations; the larger the housing stock, the higher the number of associations and the greater level of support (Cairncross *et al.*, 1990). The conclusion of the researchers is that 'the levels of assistance provided to tenants' groups was very low, and did not match the apparent commitment of councils and housing associations to tenant participation' (Cairncross *et al.*, 1989, p.55). The local authorities were not matching their rhetoric with real material support.

Put the other way round, from the tenants' point of view this meant that, although most had a meeting room and a typewriter, less than a quarter had access to office supplies or premises, training or transport facilities, and less than two-fifths had access to expert advice; those with access to genuinely independent advice were even fewer. Nearly a third were surviving on an income of below £50 per year, with membership fees and grants being the most common source of income. The conclusion is:

> Their existence is one of hand to mouth survival. In such a condition of insecurity they may find it difficult to meet the demands that arrangements for tenant participation inevitably place upon them.
>
> (Cairncross *et al.*, 1989, p. 84)

There are signs of hope, though. Where local authorities *have* put resources in, strong results have followed. Greenwich Borough Council has eight tenant support workers, including three working with black and ethnic minority groups, and has a budget of £50,000 per year for tenant groups. As a result, 78 tenants' associations and a federation now exist, and 70 per cent of tenants are represented. The lesson is that the extension and democratisation of the tenants' movement *can* be achieved, but that it costs money (Bartram, 1988).

It also requires *training for tenants*, as well as for housing workers and councillors. Yet in the above survey only two authorities provided training for tenants. Others were to some extent meeting the need: the Priority Estates Project developed tenant training on the estates on which it worked, while the national Tenant Participation Advisory Services were invited by many councils to develop participation, and also developed much-needed expertise in tenant training. The 1990s should see a more systematic training for both tenants and housing workers; an Institute of Housing National Certificate in Tenant Participation started in four colleges in England in September 1991 and is expected to spread to colleges in almost every region and country in the United Kingdom by 1992. It is designed for tenant activists and co-op members as well as for tenant liaison officers and other professionals. However, the commitment of landlords to encouraging and funding places for tenants has still to be tested.

Accountability

Beyond the consultation stage, tenants can go in one of two directions: either in the direction of making landlords more accountable, or of involvement in management. Each of these two strategies involves a hierarchy of tenant participation and control. The accountability strategy involves gaining negotiating rights, then carving out areas of choice over services, and finally being able to choose or refuse a new landlord (see Cairncross *et al.*, 1990). The involvement in management strategy involves joint management of estates, followed by self-management and finally tenant ownership of an estate. Of course, in practice they overlap, and both strategies can be pursued at once, the former in relation to the council as a whole, and the latter in relation to particular estates.

Accountability is about calling people to account for the use of powers or resources which have been lent to them on trust by others. It includes the notion of a right and also the *ability* to call to account; the one without the other makes the concept inoperative. An accountability strategy often involves tenants' groups in putting representatives on decision-making committees, and backing them up with associations and federations which mirror the landlord's decision-making machinery at all levels. The aim is to obtain negotiating rights, keeping distinct the roles and responsibilities of tenant and landlord, and forcing the latter to justify the service provided and to promise changes which suit the tenants' interest.

Tenant representation on committees was a live issue during the 1970s; between 1970 and 1973 Dick Leonard MP attempted three times to have a Council Housing (Tenants Representation) Bill passed, but without success. During the 1980s the idea of a right to representation became much more muted. There was nothing in law to stop landlords involving tenants as co-opted members of formal decision-making committees, but in practice there has been great resistance to the idea. By 1986 there were only four local authorities, all in London, which had allowed tenants on to the housing committee with full voting rights, the same number as Richardson had recorded in 1975! Two allowed partial voting rights, again the same number as in 1975. However, the number allowing tenants on the sub-committee had increased from three to twenty (Cairncross *et al.*, 1989). Further development will be inhibited, because the Local Government and Housing Act 1989 has recently prevented co-optees of housing committees and sub-committees from voting. Tenant co-optees may still serve in an advisory capacity and may gain some power from their membership but, without formal power, have no guarantee that their role will not slip back to consultation. However, the issue is once again becoming a live one in situations where the landlord wishes to transfer the housing stock to a purpose-made housing association, and tenants can demand places on the management committee as the price of their agreement.

Beyond negotiation, tenants can gain space for making choices between options. The most familiar form this takes is in modernisation schemes, where tenants often choose different types of work from a budget (see IoH/RIBA, 1988). Recently, choice has been developed in a few councils over levels of service, with tenants of tower blocks choosing, for example, whether to opt for a new concierge and pay a surcharge on the rent. During the 1980s, though, the opportunity for choices spread to include opting for a new landlord. Around the middle of the decade, some influential voices began to be heard arguing for the breakup of council housing, or at least for the right of tenants to choose a new landlord. An ex-chief housing officer, Alex Henney, argued that council housing has simply not been accountable and cannot be redeemed in its present form (1985, p. 5). He proposed the transfer of the entire council stock to Housing Management Trusts, each of 500–2,000 dwellings, which would have a majority of tenants on the board. At around the same time, the Audit Commission concluded that, unless those councils whose housing management was poor took the necessary steps to improve the service, central government might have to bypass the local authority and take over (1986, p. 82). The *Inquiry into British Housing* argued that, given purchasing power, tenants ought to be able to choose between landlords, to 'shop around', but also 'to obtain housing from a different landlord if service from the first is inadequate' (NFHA, 1985, p.8). As Clapham says, 'The right to go elsewhere is an important one to consumers, because it offers a significant element of control through choice' (1990, p. 65). Of course, the control which is gained need not just be exercised in the direction of change, because the ability to choose to end a contractual relationship which is not working satisfactorily can also act back on the relationship, giving tenants more power to renegotiate their existing situation.

One can detect elements of all these arguments in government policy as it was expressed in the 1988 Housing Act, and yet the twin strategies of Housing Action Trusts (HATs) and Tenants' Choice transfers were both, at least initially, a good deal more authoritarian than critics of council housing had expected. HATs owed more to urban development corporations than to any housing models; the aim was for a trust set up by the Secretary of State and led by business people to lever private funding into the worst estates, and over three to five years repair the stock, improve the environment and introduce diversity of tenure. Tenants would have no rights at all until the end of the HAT, when they could choose a permanent landlord. The authoritarian nature of the proposal, with its denial of any ballot among tenants and of any representation on the HAT board, and its weak promise of consultation of residents, led tenants to oppose all six of the original HATs. Then, after some vigorous campaigning, and in a last-minute amendment in the House of Lords, they eventually obtained the right to a ballot. Further concessions followed; a promise was exacted that councils would be given the necessary finance to buy the estates back at the end of the HAT's life, if the tenants so wished. Some tenants' groups opposed the HAT from the start, while others found themselves in a dilemma because of the promise of substantial government investment which, it was made clear, would not otherwise be available. They had to take up a neutral stance and to ask their tenants to make up their own minds. Yet, so deep was the suspicion that had been engendered and so worrying the remaining questions about future rent levels and security of tenure and other rights, that tenants eventually voted down all six HATs.

There may eventually be one or two HATs, but they will be formed out of a partnership of the council, tenants and central government, in which tenants will exact much stronger guarantees of rights and of representation; at Hull, where one estate is involved, and at Waltham Forest, where four estates may transfer, five out of eleven seats on the board will be reserved for tenants and councillors.

A similar pattern can be detected in the evolution of Tenants' Choice. At first, this was envisaged as a right for private landlords and housing associations to identify estates they were interested in buying. Tenants were given the right to a curious form of 'negative vote', in which those who did not vote would be counted as 'yes' voters. Unlike HATs, Tenants' Choice allowed those who voted to stay with the council to do so, but flat-dwellers would have no guarantee that their rents would not rise, as the council would have to lease their individual flat back from the new landlord. Again, the resulting controversy over voting arrangements soured relationships, and this, coupled with anxieties over the loss of rights through the new 'assured tenancies', made tenants suspicious of the whole idea; so far, no hostile bids from private landlords or associations have got off the ground. In fact, it had the opposite effect of galvanising a new tenants' movement, out of which a National Tenants and Residents Federation was formed, and it led to much bargaining with the local authorities; tenants had at last gained some leverage and could engage in genuine negotiations over the quality of services. Furthermore, the Housing Corporation, charged with registering the new landlords, has developed a set of requirements which have so far deterred private landlords. Tenant transfer staff have added into the statutory process an initial informal stage, in which tenants are able to choose between genuine alternatives, including a tenant-led buy-out.

Similar problems have arisen in relation to *new town transfers*. In England, tenants have had to choose which landlord to transfer to; they have not had the option of staying with their existing landlord (the development corporations), which are to be wound up. The government would clearly prefer transfer to a consortium of housing associations, as happened in Central Lancashire New Town. However, in some cases tenants and local authorities campaigned to allow transfer to the council; at Telford, the local authority went to the High Court, insisting that tenants had the right to a ballot, and a proposed transfer to housing associations was called off. The DoE counter-attacked by insisting that a Tenants' Choice type transfer be used in future; individual tenants would choose their new landlord from a selection which included the council and housing associations, and those who did not vote would have their homes apportioned out later between the competitors. The result at Peterborough was a huge majority in favour of the council, and as a result the DoE instructed remaining corporations to enter into management agreements with housing associations, so that tenants could get to know what they have to offer before being balloted. At Basildon, intense opposition from the tenants' action group and the local council led to such a management transfer being abandoned. At Milton Keynes, the council was sacked from a management agreement it had held after it had surveyed tenants on their preferences for management co-ops, and a transfer to a housing association was proposed. Tenants then had to threaten legal action because they had not been consulted on the transfer. As in the proposed HATs, relationships were soured and the opposite result occurred to that which the government intended; tenants are expected to vote for the council.

It is a pity that negotiations have been so clouded by the distaste of government ministers for the local authority option, though they would argue that tenants are too ready to vote for the council, not knowing enough about the alternatives. When tenants have entered more actively into the process they have obtained a good deal; at Runcorn they negotiated with a consortium of housing associations and gained a seat on management committees and a package of rights: to fair rents, the right to buy, an enhanced right to repair, and assurances over security of tenure. However, it is clear that tenants have had to fight hard to be consulted and to be offered a fair choice between alternatives.

In Scotland, the government managed to avoid the problems of tenants choosing the 'wrong' landlord, by denying altogether the right to transfer to the local authority (Enterprise and New Towns Scotland Act). Coupled with the lack of a right to consultation on major changes (the Scottish equivalent of the 1980 'Tenants Charter' did not include this), it seemed that tenants would have very little choice at all. Yet, such was the criticism from the Scottish Consumer Council, the Scottish Federation of Housing Associations, Shelter, members of Parliament and from tenants' groups, that the Scottish office had to agree to consult tenants' groups and to consider transfer to local authorities, though ministers refused to accept an amendment giving an automatic right to such a transfer. Again, the government has expressed its commitment to reducing, not adding to, the size of the public sector, and management agreements with housing associations may be entered into so that tenants can gain wider experience of landlords and have time to examine fully all the options; these will include associations, councils, private landlords and co-ops.

Yet tenants are, as in every other case of change of landlord, worried first about security of tenure, then about rent levels, and then about the standard of service; the local council is regarded as more reliable on the first two counts, if not the third. At East Kilbride, while 60 per cent of tenants surveyed said they would prefer to transfer to the council, the staff were submitting plans for a management buy-out. Such alarm was caused to tenants that the idea had to be scrapped and replaced by a new housing association, which would have tenants on the board. At Livingstone a staff agency is also planned to manage the stock, with a view to a management buy-out in 1995, but this time tenants were promised places on the board of a non-profit agency, with surpluses ploughed back into new housing. Tenants have, therefore, had a significant impact on the options being made available. Unfortunately, as in England, one result of the uncertainty is that many tenants will be panicked into exercising one right they currently have, to buy their home themselves.

One development which took the government by surprise was the very rapid development of a *voluntary transfer option* by several local councils. It was a matter of interpretation whether these were a way of pre-empting central government policies, or simply a different way of carrying them out. But, again, from the tenants' point of view the main questions concerned the extent of choice and of information made available to make a rational choice, the safeguarding of rights to security and reasonable rents, and the influence of tenants on the future landlord's decision-making structure.

First, the extent of the choices open to tenants has been questionable. Early ballots did not bode well; they aimed to predate the 1988 Housing Act, thus stopping tenants from having a choice of landlord (though they also avoided the less secure assured tenancies). Tenants were given a choice, but only to transfer to one new landlord or to stay with the council, and there was no right, if the transfer went ahead, for individuals to stay with the council if they wished. Later transfers, such as at Ryedale, were slowed down by the DoE to allow time for the Housing Corporation to insert a prior stage allowing tenants to consider first the various options available under Tenants' Choice transfers, yet in other cases, such as Bromley, tenants have still complained about being rushed into an early ballot. None of the councils pushing transfer have taken seriously the option of a tenant controlled co-operative, nor have they put forward a convincing case for remaining with the council. More generally, it has been argued that the effect of the 1989 Local Government and Housing Act on councils is to propel tenants into choosing a new landlord in order to escape steep rent rises which have been brought about by central government precisely to have that effect. Even more sinister have been the steep rent rises proposed by some councils, it is suspected, in order to close off the option of staying put, or to punish tenants who have turned down transfers. Lastly, Tenants' Choice is being ignored, for instance at Rochester, where trickle transfers of vacant property are to go ahead despite the tenants choosing to remain with the council.

Second, tenants have complained of lack of information. At Broadland, for instance, tenants complained that they only saw their proposed tenancy agreements when they received the ballot papers. When information has been given, it has sometimes been misleading, as Gloucester tenants have complained to the ombudsman, and yet tenants' groups – at Gloucester and Broadland, for example – have been denied funds to obtain their *own* advice. On the other hand, advice from interest groups arguing against transfer has also been questionable, with scaremongering over the security of assured tenancies, the lack of subsidy when stock is transferred, and the obscuring of the difference between housing associations and profit-making landlords.

Then there is concern over the type of landlord proposed. Tenants have been more inclined to reject transfers to existing large housing associations than to new, locally based ones. More seriously, some council officers – at Rochester, Rochford and the Scottish Special Housing Association, for instance – have aimed to set up private companies to take over the stock, and tenants have demanded representation and assurances about their non-profit nature. Tenants have also been concerned about the costs of the transfers, which have sometimes fallen on tenants through being charged to the housing revenue account; at Elmbridge the tenants' federation complained about this to the district auditor.

Then there is the voting method, which at first caused controversy. The Tenants' Choice-type negative vote was used at Torbay, where the majority voted against the transfer but, under the rules, it would have gone ahead; the DoE eventually stopped it. All others are now using a simple majority vote, but even this – at Broadland and Suffolk Coastal, for instance – has allowed less than half the tenants to vote in a new landlord.

On the other hand, tenants have in some areas been galvanised into action, either to oppose transfers or to obtain assurances and 'seats on the board' which show that, when given a real choice, they can gain real negotiating power. It remains to be seen whether this power can be retained over time, or whether the interests of 'workers and citizens' will once again eclipse those of the consumers.

Involvement in management

Beyond consultation, some tenants' groups, particularly at estate level, want to get involved in being responsible for management, either jointly with the landlord or through self-management. Some even take the further step of buying their estate and becoming their own landlord through a co-operative or community-based housing association, using either voluntary transfer or Tenants' Choice mechanisms.

Probably the first place *joint management* tried out was in Rochdale.

Housing management sub-committees were offered as an option for tenants on each estate, in which tenants would be the majority with councillors and officers. The sub-committee draws up an estate budget which, though it has to be approved by the council's housing committee, is then under the control of the estate committee, which can vary the expenditure within the budget. There are real benefits to the housing department from this model: it provides value for money in expenditure, and detailed feedback on the work of direct labour and other departments which can then be called to account. Around 15–20 of these Area Housing Management Sub-committees are planned. To be a success, they require a network of active tenants' associations and a strong tenants' federation, a proper training programme, as well as commitment from local councillors and officers.

A more formal type of joint management evolved out of the experience of the Priority Estates Project (PEP). In 1979 a Tenant Board was established on the Wenlock Barn estate in Hackney, similar to the Rochdale model. But there was a need to strengthen further the control of estate services by consumers and the idea of estate management boards (EMBs) emerged, designed to operate at one remove from the council via a legal management agreement (see Zipfel, 1988). There are now around ten boards being developed, both by the PEP and independently by local authorities. They are particularly suitable for larger estates, where tenant participation is not likely to be high enough to warrant a tenant management co-op, and where tenants do not want total responsibility for estate services.

Tenant management co-ops (TMCs) have been around much longer, since 1975, when councils were first allowed to pass on the management of their estates to the tenants via an agency agreement. Tenants have never had the right to self-management; the 1986 Housing and Planning Act gave them the right merely to a 'reasoned reply' from the council to their request to set one up. There are around 60 in council property. Islington has 18 with another 10 in the pipeline, while Glasgow has 13, the key to their success being a solid commitment by the local authority; in both cases co-op development units were set up in the 1970s. They have been very successful at delivering an efficient and effective service, particularly on repairs and void control (see Power, 1988; Birchall, 1988). Their growth has, however, been relatively slow, the main reason being lack of support from local authorities. However, since the 1988 Housing Act, there has been an almost total change of attitude to them; councils wishing to retain their stock have seen TMCs and EMBs as alternatives to Tenants' Choice and, in some cases, Housing Action Trusts, and many more TMCs are now being promoted. The National Federation of Housing Co-operatives has developed a modular management agreement which allows tenants to take on any combination of services, ranging from just repairs to full management including rent collection.

Tenant ownership has, at the individual level, been one of the main planks of government housing policy during the 1980s – in the shape of the right to buy. In fact, since it was granted in 1980, no other right has been so strengthened both in the level of discounts to tenants and the severity of penalties for landlords who impeded sales. Yet it is a right given not to strengthen the status of tenants but to undermine it. Collective ownership by tenants is quite different. 'Par value' ownership co-operatives have, since the mid-1970s, been a small but significant part of the social housing sector; there are about 270 of them, again concentrated in certain areas, notably London and Liverpool. After rapid growth during the late 1970s, they were affected by cuts in Housing Corporation funding, and their growth slowed down to about twenty new co-ops per year. Changes in Corporation funding since 1988, with the emphasis on the attracting of private finance, are likely further to restrict the formation of new co-ops, and the emphasis of government policy has switched more towards the development of management and ownership co-ops out of existing council stock (DoE, 1989) and of tenant participation in both this and the housing association sector (see Birchall, 1992).

Transferred ownership co-ops were first developed in the mid-1980s in Glasgow; there are six at present (including one registered as a community housing association), but many more are planned as part of a drive to transfer 25 per cent of the housing stock to other tenures. Several other councils in Scotland are also developing them, and there is the promise of Scottish homes funding for the sales and for subsequent improvement work. Under pressure from central government via the Welsh Office and 'Estate Action' funding, some Welsh councils are likely to see transfer to housing associations as the only way of raising finance for the improvement of hard-to-let estates; a tenants' association has led the way at Glyntaff Farm Estate, Pontypridd, in actively seeking such a move. In England, transfers are more likely to take place through Tenants' Choice, usually without the consent of the landlord. English Labour-controlled authorities, particularly in the southeast, are more determined to keep their stock to fulfil their homelessness responsibilities, and are generally less favourable politically towards communitarian – as opposed to municipal – socialism. Conservative authorities are more likely to be exploring total transfer of stock to a new housing association. Tenants' groups have begun the Tenants' Choice process in several cases, but for a variety of reasons. On the Walterton and Elgin estate in Westminster, they want to protect council housing from a Conservative council determined to sell vacant stock for owner-occupation. On the Trowbridge estate in Hackney, they want to generate funding for improvements to the low-rise housing, leaving the high-rise to be demolished by the council. On the Elthorne estate in Islington, three tenant management co-ops are exploring the option of going one further and becoming collective owners. It is likely, then, that a small number of transfers will be affected, but that the effect on council housing will be more to provide leverage for tenants to obtain better-quality services and greater involvement in decision-making, rather than large-scale transfer of stock to the co-operative sector. Finally, the Torbay tenants have, since the disastrous bid for voluntary transfer by the council, set up their own housing association with a view to transferring the entire stock to a tenant-controlled body. This will set an interesting precedent, particularly for new town tenants and others who have come to distrust the motives of their existing landlords.

CONCLUSION

From the tenants' point of view, the report on council housing departments during the 1980s could be phrased 'tries hard, but could do better'. The overall impression is, first, of a minority of councils who would have tried hard to give their tenants rights and improve the service, whether or not they had been pressured into doing so. But the stronger impression is of a majority, who at the beginning of the decade had to be prodded into recognising that tenants have rights at all, and then by the end of it were trying desperately either to placate the tenants with belated promises of participation and improved services, or to off-load their housing stock completely. On the other hand, the tenants' view of central government would be no less negative. A modest package of rights was granted in 1980, and some choice over the landlord in 1988, but whenever the government has found such rights and choices threatening to interfere with the over-riding aim of breaking up the council stock, it has ridden roughshod over them. More generally, the choices which tenants have now secured – mainly through their own efforts – are to be exercised in a climate of soaring rents and continuing lack of funding for estate improvements.

The underlying problem – that council tenants have not had a place in the structure of control – means that they have been prey to powerful interests whose first loyalty has lain elsewhere: to private developers, to direct labour organisations, to the local political party, and so on. The attack on, and defence of, council housing has been conducted mainly within the arena of citizen and worker interest at both local and national levels, in a policy struggle which has tended to appeal to, manipulate and redefine the consumer interest. Council tenants have been, to a large extent, pawns in a bigger game.

The 1990s will probably see three major strategies to win over tenants (see Simpson, 1989). If a Conservative government is re-elected, it might simply wait for the combined effects of its 1988 and 1989 legislation to come to fruition. Had this paper been written a year ago, it would have reported the probable frustration of the government's plans by tenants voting against HATs, and councils promising a new deal for their tenants. But the 'carrot' of tenants' transfer, followed by the 'stick' of rapidly rising rents and drastic restrictions on capital spending, have already, early in 1991, produced an intensification of plans for voluntary and Tenants' Choice transfers, along with the individual right to buy. However, the government has at least one more trick up its collective sleeve; along with the rents-to-mortgages scheme which may or may not work, it will extend compulsory competitive tendering to council housing management. This would have the same effect as did Tenants' Choice, of raising the spectre of private landlords taking over the stock, and it would force councils to adopt an agency structure – already being experimented with in a few authorities – which would finally separate housing management from other services and make it possible for tenants to be represented 'on the board'. It would also give an added impetus to tenant management co-ops and estate management boards, which might well want to bid under any private-sector competition in order to retain control over estates. Like Tenants' Choice, it would make councils and their tenants do 'the right thing for the wrong reasons', but without any guarantee of having the resources to do it really well.

On the other hand, the election of a Labour government would bring in a new partnership of central and local government, which would probably, along with a modest de-restriction of capital spending and increase in revenue subsidy, emphasise the need to improve quality in housing management: the formation of an advisory inspection service (probably based on the DoE or perhaps the Housing Corporation) to monitor performance in housing management, stricter management accounting for services, a stress on staff training, and a new tenants' charter to give enforceable participation rights and a mandatory framework for resourcing tenants' organisations (Simpson, 1989). This time around, the local authority associations and the Institute of Housing would be only too glad to support such a tenants' charter (see Institute of Housing, 1989 and 1990; Association of London Authorities, 1988).

A coalition of Liberal Democrats and either of the two main parties might lead to that most radical of plans, suggested by some commentators on both right (Henney, 1985) and left (Clapham, 1989): the transfer of all council housing to tenant-controlled bodies. Given a sensible and attractive framework for its financing, a positive strategic and monitoring role for the local authority, and some kind of transfer commission to promote and supervise the change-over, it *could* provide a positive libertarian yet socially just framework for the de-municipalisation of a service which probably should never have been municipalised in the first place.

REFERENCES

Association of London Authorities (ALA) (1988) *Tenants in Power*, London: ALA.
Audit Commission (1986) *Managing the Crisis in Council Housing*, London: HMSO.
Bartram, M. (1988) *Consulting Tenants: Council Initiatives in the Late 1980s*, London: Community Rights Project.
Birchall, J. (1988) *Building Communities: The Co-operative Way*, London: Routledge & Kegan Paul.
——(1992) 'Housing Co-operatives in Britain', Department of Government working paper no. 21, Uxbridge: Brunel University.
Cairncross, L., Clapham, D. and Goodlad, R. (1989) *Tenant Participation in Housing Management*, London: Institute of Housing/Tenant Participation Advisory Service.
——(1990) *The Pattern of Tenant Participation in Council Housing Management*, Discussion Paper no. 31, Glasgow: Centre for Housing Research.
Clapham, D. (1989) *Goodbye Council Housing?*, London: Unwin Hyman.
——(1990) 'Housing', in Deakin, N. and Wright, A. (eds) *Consuming Public Services*, London: Routledge.
Clapham, D., Kemp, P. and Smith, S. (1990) *Housing and Social Policy*, London: Macmillan.
Cole, G.D.H. (1980) *Guild Socialism Restated*, New Brunswick: Transaction.
Community Development Housing Group (1986) *But Will It Fly, Mr Wright?*, Glasgow: Tenant Participation Advisory Service.
Day, P. and Klein, R. (1987) *Accountabilities: Five Public Services*, London: Tavistock.
Deakin, N. and Wright, A. (eds) (1990) *Consuming Public Services*, London: Routledge.
Department of the Environment (DoE) (1989) *Tenants in the Lead: the Housing Co-operative Review*, London: HMSO.
——(1990) *Reports to Tenants etc. Determinations*, London: HMSO.
Fox, D. (1973) *Conditions of Tenancy: a Possible Alternative Approach*, London: HMSO.
Goodlad, R. (1986) *Telling the Tenants*, Glasgow: Scottish Consumer Council.
Henney, A. (1985) *Trust the Tenant: Devolving Municipal Housing*, London: Centre for Policy Studies.
Housing Corporation (1989) *Performance Expectations: Housing Association Guide to Self Monitoring*, London: Housing Corporation.
Housing Services Advisory Group (1978) *Allocation of Council Housing*, London: Department of the Environment.
Institute of Housing (IoH) (1989) *A Tenants Charter for the 1990s*, London: IoH.
——(1990) *Social Housing in the 1990s: Challenges, Choices and Change*, London: IoH.

Institute of Housing/Royal Institute of British Architects (IoH/RIBA) (1988) *Tenant Participation in Housing Design*, London: IoH/RIBA.

Kay, A., Legg, C. and Foot, J. (1986) *The 1980 Tenants Rights in Practice*, London: City University.

Klein, R. (1984) 'The politics of participation' in Maxwell, R. and Weaver, N. (eds) *Public Participation in Health*, London: Kings Fund.

Laffin, M. (1986) *Professionalism and Policy: the Role of the Professions in the Central-Local Government Relationship*, Aldershot: Avebury.

London Housing News (1990) July.

Maclennan, D., Clapham, D., Goodlad, M., Kemp, P., Malcolm, J., Satsangi, M., Stanforth, J. and Whitefield, L. (1989) *The Nature and Effectiveness of Housing Management in England*, London: HMSO.

Malpass, P. (1990) *Reshaping Housing Policy: Subsidies, Rents and Residualisation*, London: Routledge.

National Consumer Council (NCC) (1976) *Tenancy Agreements Between Councils and their Tenants*, London: NCC.

National Federation of Housing Associations (NFHA) (1985) *Inquiry into British Housing*, London: NFHA.

——(1987) *Standards for Housing Management*, London: NFHA.

Platt, S., with Newton, C. and Willson, M. (1990) *Whose Home? – Accountability to Tenants and Communities*, London: NFHA.

Power, A. (1988) *Under New Management: the Experience of 13 Islington Tenant Management Co-ops*, London: Priority Estates Project.

Richardson, A. (1977) *Tenant Participation in Council Housing Management*, London: DoE.

Simpson, D. (1989) 'Future postponed', *Roof*, March–April.

Ward, C. (1974) *Tenants Take Over*, London, Architectural Press.

Webb, S. (1891) *The London Programme*, London: Swan Sonnenschein.

Webb, S. and Webb, B. (1930) *The Co-operative Movement*, London: Longman Green & Co.

Zipfel, T. (1988) *Estate Management Boards*, London: Priority Estates Project.

Conclusion

Johnston Birchall

The idea of a crossroads implies not only that there is a need to make a clear choice about the direction of policy, but that there is also room to manoeuvre, to change direction. Political parties all seem intent on making clear their differences over housing policy, and their determination to go in different policy directions; if they look back over their shoulders, they might just see the old 'Butskellite consensus', but it may be too far back down the road even to be recalled. Yet, if we change the metaphor, we can see that choice is a luxury which, in some areas, policy-makers can no longer afford.

Peter Malpass likens housing policy to a supertanker which cannot easily slow down or change direction. Supposing we turn his metaphor into an allegory, and visualise it more as a flotilla of ships negotiating their way through a tidal estuary. The larger ones are the more important policies. The policy towards owner-occupation is a huge supertanker which cannot easily be manoeuvred or slowed down, for two reasons. First, any delay in its passage, any change in the investors' expectations of where it is going, may be economically too costly. Second, imagine the estuary it has to navigate as a constantly shifting accumulation of sediment which might represent the changing housing stock, its condition and pattern of ownership and control. The more owner-occupation becomes the norm for households, the narrower the channel becomes, and some directions which had been navigable a few years ago become silted up. This is why the party policies towards it are now so cautious. Hardly anyone outside the academic ivory tower (or lighthouse, since we spend so much time warning of disaster?) will dare to attack mortgage interest tax relief for ordinary rate taxpayers, and the discussion over the introduction of an imputed rental income tax, even over capital gains tax, remains a kind of academic board game, useful for demonstrating our skill and passing the time while someone else steers the ship.

Yet if we imagine the flow of ships down the estuary as, in the image familiar to poets and sages, part of the flow of time, we can see that the possibilities for change of direction and pace are themselves always changing; nothing is fixed, at least not for more than a decade or two which, in the run of things, is not very long. Like the supertankers, which seemed such a marvellous economic advance a decade ago but which, if they run aground or catch fire, have such disastrous consequences for the environment, the policy of continually expanding owner-occupation carries high risks. The growing problem of disrepair in the sector is well known, but hardly anyone would have predicted the severity of the current recession in the housing market, nor the crisis which high interest rates would bring to a politically significant minority of citizens, a much larger minority than was originally encompassed by the term 'home ownership at the margins'. Yet the response to the crisis in debt and repossession, stagnation in the property market and inhibited labour mobility has been . . .what? The crisis shows how little room for manoeuvre there is, even for a government single-mindedly committed to the tenure.

It may be that, as the cost of mortgage interest tax relief declines, some of it, or some extra reliefs, might be targetable at first-time buyers. Arrangements might be made for shared ownership for those caught with mortgage debt. But Smallwood has shown just how constrained the building societies also are by their new, fiercely competitive environment. The fact that such simple and attractive measures have to be couched in such conditional terms shows just how narrow the 'deep water channel' of policy has become. The fact that they are espoused by the opposition parties shows that all they can do is try to out-do the government with measures which further accelerate the supertanker. The political imperative in a majoritarian democracy is to keep the majority satisfied. As Lowe shows in his chapter, the direct accumulation of rewards from ownership will be relied on more and more by citizens during the 1990s, to enable their children to be housed, their elderly relatives to be cared for, and their private medical care afforded. Those who have not invested in what the supertanker is carrying will be severely disadvantaged, but will also be politically marginal. Those who have invested heavily will be increasingly anxious about an investment which governments seem unable to protect; the pilot of the policy 'supertanker' may always have to give way to economic policy.

Council housing policy is a smaller vessel, almost as difficult to manoeuvre, because the channel it is navigating has also been becoming more and more dangerous (though in this case more by deliberate neglect by the harbour authorities than by natural causes!). Malpass' analysis shows that this vessel also has been accelerating in one direction over the past decade. Yet here there are more choices because, unlike home ownership whose promotion has involved positive tax concessions, measures to inhibit the natural advantages of council housing (rent pooling, its non-profit nature, leading to low rents) have been mainly negative. This means that, even without the injection of new public funding, councils can gain more room for manoeuvre simply by being allowed to do what they want, and already have the resources, to do: using capital receipts to fund major repairs, borrowing on the open market to build new houses, entering freely into partnership deals, and so on. On the other hand, much of what has been achieved by the Conservatives cannot be undone; all that opposition parties can do in relation to the more libertarian aspects of policy is to offer to accelerate the vessel in the same direction, offering the right to buy to assured housing association tenants, giving Tenants' Choice to housing associations and even private tenants.

The problem for defenders of council housing as a tenure is that it really is at a crossroads, in this sense – another term of Conservative government will probably see the final breakup of the stock, simply because the measures already put in place will have had time to take effect. For many, voluntary transfer to housing associations seems the only alternative. Those local authorities who want to fight to retain their stock have a real problem of credibility. Birchall's analysis of councils during the 1980s shows that there is still a long way to go before they can be said to be good landlords. Yet the process of changing landlords is such a protracted and risky one, with real problems of access to unbiased information and uncertainty about alternatives, that tenants may be tempted to opt for 'the devil they know', so the question of the quality of council housing will remain high on the policy agenda.

As Langstaff has shown, the housing association movement experienced a relatively protected environment during the 1980s, but is now definitely at the crossroads. Expansion of the top twenty associations may mean that they expand more quickly during the 1990s, but this may be at the expense of the movement's 'soul'. Those organisations which have made it such an interesting tenure – housing co-ops, community-based housing associations, special needs groups – are already in difficulties with the new funding arrangements, and will have to form consortia, or give up their development to the large agencies, if they are to survive. If Birchall's analysis had been continued to include housing associations, it would have been found that they are not much better than councils in the way they treat their tenants; in fact, their practice is much worse than that of the most progressive councils. A new housing inspectorate, charged with monitoring all forms of social housing, would be a significant change of policy direction, achievable for very little cost, but in relation to council housing, the government is putting its faith in the much more blunt instrument of compulsory competitive tendering.

If policy towards owner-occupation and social rented housing are supertankers which have, during the 1980s, been steaming steadily in one direction, policy towards the private rented sector has, as Crook shows, put to sea hesitantly in 1980 and then gathered steam for a record-breaking run from 1988. Looking further back, we can see it as an erratically zigzagging vessel, lurching from rent control to deregulation and back ever since 1915, and, as Coleman emphasises, in consequence being regarded as a very unsafe investment indeed. To change the metaphor, the private rented sector may well be at a critical crossroads during the early 1990s. It was at a legislative crossroads in 1988, but such is the slow pace of change, with most tenancies still being regulated, that the effects will take much longer to show up; indeed, as Crook argues, the effects may be on a small scale, and so could easily be reversible by a change of government. This is, as Coleman sees it, the dilemma for investors. On the other hand, continuation of the Conservative government into another term could mean that housing associations become more like their private-sector poor relations, being allowed to opt for limited company status. Private landlords could, if Coleman has his way, be made eligible for the same tax concessions and grant aid as the associations. The combined effect of such measures would be to lower the status of associations in the opinion of their tenants, and raise those of the private sector in the opinion of the investors.

What of the housing stocks and tenure patterns within which housing policy operates, which we have imagined as the sand banks between which the policies have to navigate? These grow, shrink, change their shape just as policies do, but much more slowly (as the poet Norman Nicholson used to point out, even the rocks flow down to the sea). It is in the nature of housing that what is built endures through the lifetimes of several occupiers and many governments. What is allowed to become unfit for habitation, or is not built at all because of economic recession, central government restrictions on borrowing, or the lack of purchasing power, is simply a lost opportunity which reverberates down through the generations. That is why the governments of the 1980s and 1990s will ultimately be judged not on the extent to which they accelerated or changed direction with particular policies towards different tenures, but on the extent to which they encouraged the production, conservation and maximum use-value of the housing stock.

Index